Wireless Local
Area Networks

Other Books in the McGraw-Hill Computer Communications Series

BLACK • *TCP/IP & Related Protocols—Second Edition*

BLACK • *Network Management Standards: SNMP, CMIP, TMN, MIBs, and Object Libraries—Second Edition*

BLACK • *The V Series Recommendations: Standards for Data Communications over the Telephone Network—Second Edition*

BLACK • *The X Series Recommendations: Standards for Data Communications—Second Edition*

HUTCHINSON • *ISDN Computer Applications Development*

NEMZOW • *Implementing Wireless Networks*

PETERSON • *TCP/IP Networking: A Guide to the IBM Environment*

SCHATT • *Linking LANs—Second Edition*

SIGNORE, CREAMER, STEGMAN • *The ODBC Solution: Open Database Connectivity in Distributed Environments*

NEMZOW • *Enterprise Network Performance Optimization*

FORTINO, RYBACZYK • *Network Administration 101*

In order to receive additional information on these or any other McGraw-Hill titles, in the United States please call 1-800-822-8158. In other countries, contact your local McGraw-Hill representative.

Wireless Local Area Networks

Technology, Issues, and Strategies

Peter T. Davis

Craig R. McGuffin

McGraw-Hill, Inc.

New York San Francisco Washington, D.C. Auckland Bogotá
Caracas Lisbon London Madrid Mexico City Milan
Montreal New Delhi San Juan Singapore
Sydney Tokyo Toronto

Library of Congress Cataloging-in-Publication Data
Davis, Peter T.
  Wireless local area networks : technology, issues, and strategies
/ by Peter T. Davis and Craig R. McGuffin.
     p.   cm.
  Includes bibliographical references and index.
  ISBN 0-07-015839-8
  1. Local area networks (Computer networks)   2. Wireless
communication systems.   I. McGuffin, Craig R.   II. Title.
TK5105.7.D388   1994
004.6'8—dc20   94-34083
                                                      CIP

2 3 4 5 6 7 8 9 0   DOC/DOC   9 8 7 6 5

ISBN 0-07-015839-8

*The editors of this book were Brad Schepp and Sally Anne Glover and the
production supervisor was Katherine G. Brown. This book was set in ITC
Century Light. It was composed by TAB Books.*

Printed and bound by R.R. Donnelley & Sons, Crawfordsville, Indiana

MH95
0158398

To Nancy and Donald, unrivaled siblings, for your generous gifts of love, knowledge, guidance, understanding, and patience, and for all the happy memories.

Peter T. Davis

To the memory of Charlotte, for all your years of love and support.

Craig R. McGuffin

Contents

Preface   xi

Acknowledgments   xv

Part 1   The Concepts

Chapter 1. Changing the Network Paradigm   3

What Is the Wireless Data Marketplace?   4
A Wave Shift   8
Why Wireless Communications?   11
Wireless Data Technologies   14
  Spread-spectrum radio   15
  Infrared communications   16
  Cellular telephone modems and adapters   16
  Mobile radio   17
  Meteor burst communications   18
  Microwave   18
  Mobile satellite communications and
    Very Small Aperture Terminal (VSAT) technology   19
  FM squared   21
  FM sideband   22
Wireless Local Area Network Applications   22
Future Trends and Issues   23
  What about standards?   23
  Health and safety   24
  Security   24

Chapter 2. Introduction to Local Area Networks   25

What Is a Local Area Network?   26
Local Area Network Nodes   27
  The network interface card   28
  Network roles   28
  Services offered by nodes   29
  Client nodes   32

Connecting Media   32
  Network topologies   32
  Controlling access to the medium   34
Network Communication Protocols   36
Internetworking Devices   38
How Does Wireless Fit In?   39

Chapter 3. Wireless Technology—Its Application to LANs   41

Basics of Wireless Communication   41
  Key components   43
More on Electromagnetic Radiation   44
  Range of transmission   46
  Information-carrying capacity   47
  The need for regulation   47
  Equipment costs   48
  Adding the information—modulation   49
  Sharing the spectrum   49
Common Frequencies Used for Wireless LANs   51
Wireless Architectures   52
  Replacement of a point-to-point link   53
  Connection of wireless nodes to a wired LAN   54
  Stand-alone wireless LAN   55

Chapter 4. Standards for Wireless LANs   59

The Need for Standards   60
Where the Standards Apply   61
  The affected OSI layers   62
  Effect on upper layers of OSI Model   63
The Current Standards   64
  IEEE 802.11 standard for wireless LANs   65
  The European Telecommunication Standards Institute   67
Future Standards Development   67
The Impact of Standards on Current Wireless LANs   68

Part 2   The Needs

Chapter 5. Uses for Wireless LANs   73

Comparing Wires with Wireless   73
  Performance   74
  Cost   74
Integrating the Technologies   75
  Where wireless LANs make sense   77
  Where wiring cannot be used   77
  Where wireless is superior   79
Temporary Locations   80
Greater Mobility   81

Part 3 The Applications

Chapter 6. Wireless Applications   85

Industrial Applications   88
  Petroleum industry   89
  Agriculture and food   89
  Factory control   89
Transportation Industry   92
Office and Similar Applications   93
Wireless Disaster Recovery Applications   98
Wireless Considerations   99

Chapter 7. Industry Applications   101

Health Care Industry   101
  Health care wireless techniques   102
  Wireless benefits in health care   106
  Wireless issues in health care   106
Financial Services Industry   107
  Financial services wireless techniques   108
  Wireless benefits in the financial services industry 108
  Wireless issues in financial services   109
Services Industry   110
  Food services industry   110
  Food services wireless techniques   111
  Wireless benefits in food services   111
  Wireless issues in the food service industry   111
Retail Industry   112
  Retail wireless techniques   112
  Wireless benefits in retailing   114
  Wireless issues in the retail industry   114
  Wireless techniques in warehousing   115
  Wireless benefits in warehousing   115
  Wireless issues in warehousing   115

Part 4 The Problems

Chapter 8. Health and Safety   121

Our Electrical Environment   121
Imagine a World without electricity   122
The Invisible World   122
Measuring Electricity—Volts, Amps and Watts   123
EMF Emitters   123
Electromagnetic Fields = Electrical Fields + Magnetic Fields   124
The Electromagnetic Spectrum   124
Your Body Is an Antenna   125
Be Proactive, Not Reactive   126

Chapter 9. The Security of Wireless LANs   127

  The Need for Security   129
    Information assets   130
    Security goals   130
    Threats or risks   131
    Security measures   132
  Are Wireless LANs Secure?   132
  Wireless LAN Security Exposures   134
    Theft of equipment   135
    Eavesdropping/interception of messages   135
    Modification/substitution of data   136
    Masquerade   138
    Interference/jamming   139
  Security Measures   140
    Shielding radiation   140
    Security features in network operating systems   141
    Encryption   142

Chapter 10. Implementing and Managing Wireless LANs   147

  Implementation   147
    Pilot projects   148
    Performing the full implementation   150
  Ongoing Management   153
    Fault management   153
    Performance management   154
    Asset management   155

Part 5   The Future

Chapter 11. Future Wireless Networks   159

  The Failure of Wireless Technology   160
  The Promise of Wireless Technology   160
  Think Creatively   162

Appendices

  A. The Vendors   167
  B. Wireless Vendors   207
  C. Frequency Bands   215
  D. Organizations   217
  E. Where to Get Information Updates   221
  Glossary   229
  Selected Bibliography   239
  Index   247
  About the Authors   251

Preface

When we originally proposed writing a book, our intention was to cover the subject of wireless data communications. Even though this would have been a worthy topic, we felt that we could make a greater contribution by limiting the scope to wireless local area networks. There are a number of compelling reasons for this.

First, we believe that today's wireless technology is only the first wave of wireless local area network products. Until recently, the wireless local area network marketplace was a lot like teenage sex when we were kids. Everybody said they were doing it, but very few actually were. And those who were doing it experienced different levels of success and satisfaction. But, in the last six months we have seen the genesis of the future wireless world. In truth, it is our belief that today's technology is akin to a microwave with a great Tsunami right behind it.

At a recent COMDEX in Toronto, more than 10 percent of the local area networking vendors were displaying infrared, microwave, and spread-spectrum products. And this figure probably doubles that from the previous year's COMDEX. Even though the marketplace is in its infancy, we see many organizations transforming their businesses through the use of wireless, especially with local area networking. For this reason, we have included several chapters to start you thinking about how to use this technology to break the existing paradigms in your organization. If you aren't working on wireless today, then you might get pulled under by the coming tidal wave, and then it will be sink or swim. This book will show you how to jump in and feel comfortable with wireless LAN technology.

Second, we could not find any books that addressed only wireless LANs. Sure, there are books on wireless data communications with the obligatory chapter on LAN technology. McGraw-Hill even has a book on wireless data communications, but its focus is primarily on wide area technology (see "Selected Bibliography"). Other books focus on either cellular digital packet data (CDPD) or packet radio technology. While both are excellent technologies that will find a niche in the wireless world, they are also regulated technologies. So we foremost wanted to provide a book about a wireless technology that you can buy, implement, and try without committing to a great expense. Of course, there also are books on local area networks, but these books tend to pay attention only to standard cabling systems such as copper and fiber optic. There just were not any books dealing with wireless LANs until now.

Third, wireless computing, or mobile computing, is gaining popularity. We've seen a phenomenal growth in product offerings and wireless applications in the short time we researched and wrote this book. Organizations are embracing the wireless wide area networking technology like cellular digital packet data and packet radio. But these technologies are regulated and slow. Wireless LAN technology provides an excellent alternative for mobile computing within your organization. So we decided to focus on this one area that is rapidly gaining in popularity in many organizations.

Reading the Book

We intended this book for those individuals who are exploring wireless local area networking and looking for straightforward help. It doesn't matter whether you are a user, designer of a new network, the manager of an old network, the network administrator, the system administrator, the security administrator, the system auditor, the owner of the network, or a senior manager who wants to understand wireless technology. If you are interested in wireless local area networking, this book is for you.

We aimed this book at a wide audience because we believe in the need to proselytize network administrators and their users on the evolving technology. Everything needed for a basic understanding of the issues is included, so you need not be an expert to get the most from this book.

Organization of the Book

Material in this book has been organized to lead you through the process of selecting a wireless technology and a potential application. We encourage you to read the book in sequential order because chapters tend to build on each other. For instance, we introduce local area networking and wireless concepts early on in the book.

Part 1: The concepts

In Part 1, we discuss background issues. We describe local area networking, the wireless technologies, and the emerging standards.

Chapter 1, "Changing the Network Paradigm," explains the three waves in information processing and how wireless networks have facilitated the third wave. In addition, the chapter covers the background and the rationale for wireless LAN technology. You'll also be introduced to the different wireless data communication methods, including those for wide area networking.

Chapter 2, "Introduction to Local Area Networks" describes the basic components of a LAN, including the role of various kinds of nodes, connecting media, and network protocols. It provides a basic vocabulary for the rest of the book.

Chapter 3, "Wireless Technology—Its Application to LANs" describes the basic concepts behind wireless communications, including the key components needed to communicate and the techniques used to send data using electromagnetic radiation. This chapter also illustrates how various wireless architectures are used to build LANs.

In chapter 4, "Standards for Wireless LANs," you'll learn about the role of standards in the development of wireless LANs and the current initiatives underway in North America and Europe.

Part 2: The needs

In Part 2, we explore what requirements exist in today's organizations and how these can be met by using wireless LANs. Chapter 5, "Uses for Wireless LANs," describes some of the applications where using wireless technology makes good sense. You'll see examples of wireless LANs in action to solve common problems faced by organizations when they need to implement a network.

Part 3: The applications

In Part 3, we turn our attention to applications for your wireless local area network. You'll learn how other companies have used the technology.

Chapter 6, "Wireless Applications," provides an overview of wireless applications for the following areas: petroleum, agriculture, transportation, factory and plant control, the office, and disaster recovery.

Chapter 7, "Industry Applications," provides specific solutions for the health care, financial, service, and retail industries.

Part 4: The problems

Chapter 8, "Health and Safety," provides a look at the issues associated with electromagnetic fields. The chapter provides a brief description of electrical concepts such as volts, amps, electric fields, magnetic fields, and conductors.

Chapter 9, "The Security of Wireless LANs," deals with the confidentiality, availability, and integrity concerns that arise when using a wireless LAN.

You'll see the unique security exposures posed by wireless, as well as the techniques available to address them.

Chapter 10, "Implementing and Managing Wireless LANs," provides suggestions on getting your wireless LAN up and running in the first place, as well as keeping it running at peak efficiency and effectiveness.

Part 5: The future

Because network solutions are not static, we have provided in Part 5 our insight into how to think about implementing wireless LANs in your organization.

Chapter 11, "Future Wireless Networks," looks at some issues to be faced by organizations with local area networks in the not-so-distant future. You'll learn how to search for breakthrough applications.

Appendices

We have provided several useful appendices to supplement the information in the book's chapters. Appendix A, "Survey of Current Products," provides product and contact information for wireless LAN products and can serve as a useful starting point for your review of LAN technology.

Appendix B, "Summary of Vendors and Products," classifies products into network adapters, bridges, hubs, and concentrators. You might want to use this appendix as a quick reference.

Appendix C, "Frequency Bands," provides the current FCC breakdown of the electromagnetic spectrum.

Appendix D, "Organizations," lists those groups who can further help in your understanding of wireless data communications. Hopefully, these groups can help take over where this book leaves off.

Back Matter

We have included some other beneficial information to assist in your understanding of wireless LANs, standards, security, EMFs, and local area networking.

"Selected Bibliography" is a comprehensive list of books, manuals, publications, periodicals, and articles for wireless local area networks.

The glossary contains definitions of the major wireless, networking, and information-processing terms used throughout the book.

We hope you learn as much from reading this book as we did in writing it. If you have any comments about the book or wireless itself, please send us mail at:

Peter T. Davis 72734.36@compuserve.com

Craig R. McGuffin CRMcGuffin@CRMcG.COM

Acknowledgments

A special thanks to:

- Brad Schepp for suggesting the book and for showing great patience.
- Stacey Spurlock for administrative help with this book.
- Everyone at McGraw-Hill who helped with the finished product.
- Ann Marie Amaro, Matt Aver, Bud Bates, Michael J. Berman, Ron Boninger, Chris Carroll, Ian Cheong, Fouad Choucair, Sharon M. Cullina, Brian Davies, Scott Farley, Lisa Freeland, David E. Glidden, Steve Henshaw, Nellie Lee, John M. Levis, Arlene Lightford, John Moorhead, Robert Morris, Michael D. Olson, Patrick J. Papacek, Kenneth R. Pedigo, John "Jocko" Tannahill, Sheldon Sacks, Robyn Webber, John Whitinger, Hatim Zaghloul, and Richard "Koz" Korzeniewski for material used as background for this book.
- Everybody who encouraged us to write the book.

Peter would like to thank Craig for helping with this book and with *Teach Yourself NetWare in 14 Days*. Even though there were times when we both wondered why we did it, all things being said and done, I am sure you are as happy as I am with our work. Peter also would like to thank Janet and Ruth for their encouragement, and Kelly for her indifferent attitude to his writing. Hopefully, some day you will read this and smile.

Of course, Craig would also like to thank Peter for all his help, and for talking me into this, our second book together. Didn't you learn after the first one? I would also like to thank Adele Pugliese and Barry Lewis for their support, encouragement and exhortations to "get on with it." And to all my friends and colleagues, now and as always, many thanks for your kind understanding and guidance.

The Concepts

"The longer I live, the more keenly I feel that whatever was good enough for our fathers is not good enough for us."

Oscar Wilde

"When people shake their heads because we are living in a restless age, ask them how they would like to live in a stationary one and do without change."

George Bernard Shaw

"Many ideas grow better when transplanted into another mind than in the one where they sprung up."

Oliver Wendell Holmes

1

Changing the Networking Paradigm

Attention network shoppers. There's something new in the air. The signals are all around us. Everywhere you turn, people are talking about wireless communications. You can't pick up a paper or periodical and not read about a merger, announcement, or pronouncement regarding wireless communications. A case in point: Ziff Publications announced "ZiffNet Unplugged: A Wireless Event" in the "What's New This Week" (March 31, 1994) section of CompuServe Information Services. Host Chris DeVoney and columnist Jesse Berst were traveling around Seattle using wireless data modems to log onto ZiffNet's Executives Online Forum to chat live about wireless data transmission.

During the same month, Mobile Telecommunication Technologies Corporation (Mtel), the largest paging company in the United States, disclosed its joint venture with Microsoft Corporation to develop a wireless communications network. The venture, called Nationwide Wireless Network or NWN, will offer two-way messaging through portable devices, including laptop computers, pagers, and pocket data communicators. Along with the Mtel deal, Microsoft declared other electronic superhighway deals they had made. Microsoft announced alliances with Nippon Telegraph & Telephone Corporation and TeleCommunications, Inc. Bill Gates had a busy month networking! He wants to be part of the coming electronic highway and is betting on it as a significant growth sector for Microsoft. And why shouldn't he? Industry researchers are backing his gamble.

What Is the Wireless Data Marketplace?

Researchers suggest that the wireless data market will explode in the next few years. Forrester Research, for instance, predicts 2.6 million mobile data users by the year 1996. Arthur D. Little suggests there will be 3.8 million mobile users by 1999. And Booz-Allen & Hamilton, believing that there will be a large demand for small, hand-held personal communicators, predicts that 12 to 15 million subscribers will use public, mobile data services by the year 2000. The growth is not just in mobile workers but also in the wireless office, which is currently manifesting itself in LANs.

The wireless office is currently a small market but will grow in this decade, according to a report by Market Intelligence of Mountain View, California. In *The Wireless Office: LAN, PBX Users Cut Cables, Eye Integration*, they predict that the U.S. wireless office market will approach $500 million in 1993, $1 billion in 1994, and $3 billion by 1997.

This book explores one burgeoning segment of the wireless communications and office market—wireless local area networks (LANs). As an enabling technology, in the near future wireless LANs will augment, not replace, wirebound LANs. Wireless LANs provide unique advantages such as fast and easy installation, a high degree of user mobility, and the ability to recover your investment each time a move occurs. Advantages such as these have led industry experts to project remarkable growth in the wireless LAN market. According to a 1993 BIS Strategic Decisions study, wireless LANs will capture a healthy 17 percent of total LAN shipments by 1997. As Figure 1.1 shows, studies predict that the market will approach $300 million in this time. This would represent a major shift from current trends. Wireless LANs in early 1993 only accounted for less than ½ percent of the total installed LAN connection base. Future growth will result from improving prices compared to wirebound LANs, standards establishment, and the emergence of a critical mass of mobile computing devices. In Windata Inc.'s "Providing the Industry's First Wireless LAN to Fully Deliver the Features and Benefits of Wired LANs," they predict that the market will surpass the $500 million mark by the mid 1990s! Figure 1.2 shows the rapid growth predicted by Windata.

The progenitors of tomorrow's personal communications systems—today's cellular and wireless data networks—are, of course, already here. But while cellular is widely known and accepted in the marketplace, wireless data networks only now are gaining market acceptance. During the 1990s, market recognition is expected to change dramatically. According to a report entitled "Wireless Data Communications Markets" issued in late 1993, International Resource Development expects the U.S. market for wireless data communications and equipment to grow enormously during the next 10 years. From Figure 1.3, you can see that the market will exceed $7 billion by 1998 for wireless data communication equipment, and $240 million for wireless LAN equipment.

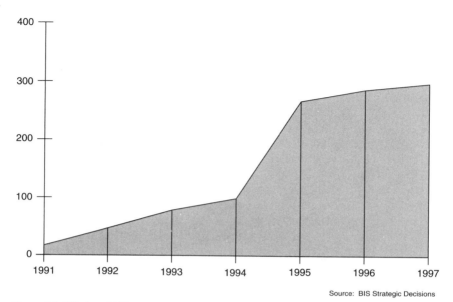

Source: BIS Strategic Decisions

Figure 1.1 Wireless LAN revenues.

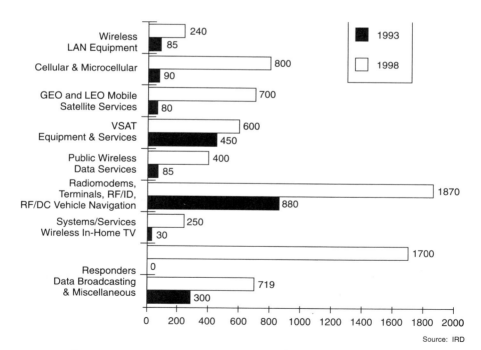

Source: IRD

Figure 1.2 Projected U.S. wireless growth.

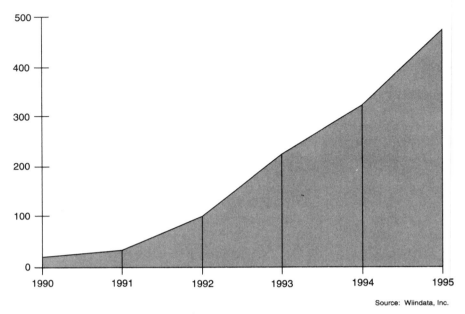

Figure 1.3 Wireless LAN market growth.

You repeatedly find these themes in other studies. Touche Ross (part of Deloitte Touche Tohmatsu International) surveyed senior executives of the top 2000 U.K. companies, and the key findings were:

- Companies expect cellular telephones, mobile data, and mobile fax to grow dramatically during the next five years.

- Enhanced customer service and increased productivity are the key benefits.

- For many companies, cost is a barrier: 90 percent placed it amongst the top four reasons for nonuse. But costs are expected to fall, with over 60 percent believing that the premium over wired services will be 10 percent or less within three years. In addition, these companies believed the technology to be poor. For example, data rates are usually less than 2 Mbps, which cannot support newer technologies such as multimedia and client/server computing.

- Benefits have been considerable for those companies that implemented wireless office systems. Not only can office layouts be planned with much greater flexibility, but the ability to contact employees easily, wherever they are, can markedly improve a company's responsiveness.

- Most executives believe that wireless communications will supplement wired communications rather than replace them. Wireless will be developed on an application-by-application basis rather than as a core facility.

- Respondents gave cost-effectiveness and reliability as the reasons for choosing wired in preference to wireless. Where companies preferred wireless communications, the reasons given were flexibility, convenience, and greater efficiency.

Another Deloitte Touche Tohmatsu International study, *Welcome to the Untethered World* (1993), asked senior decision makers in 65 medium to large Canadian companies about wireless. The researchers found the benefits and concerns for wireless LANs as shown in Figure 1.4. The Canadian study confirms the conclusions from their U.K. study.

Perhaps the experts don't agree on the magnitude of the growth, but they do agree on one thing: wireless data communications is growing. When wireless first emerged, everybody wondered whether it was a niche market or whether the market would grow to challenge wired LANs. Apparently, industry watchers and researchers think wireless LANs will challenge. They attributed the slow initial growth partly to the low capacity of wireless, when compared to the wired competition. As evidenced by Touche Ross' survey, the impetus for future growth will come from less expensive products, looming standards, and the growing mobile communications market. Until then, the technology pioneers will select wireless LANs for more compelling reasons—because they frequently move their LAN or they cannot cable the building. You find the technology pioneers working in trading floors, emergency medical, point-of-sale, factory floor, material handling and warehousing, fast food and restaurant service, security and alarms, vending, and car rental.

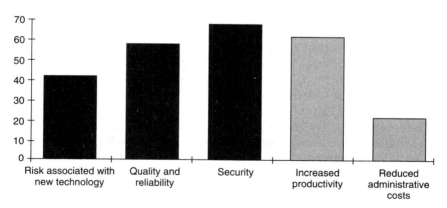

Source: Deloitte Touche Tohmatsu, "Welcome to the Untethered World"

Figure 1.4 Using wireless technology.

To understand where the wireless market is going, you need to understand where computing has been. Undeniably, wireless is a new wave in computing. What are some of the signs of this new wave in computing? What is the new wave?

A Wave Shift

Many authors have written about paradigm shifts or waves, most notably Alvin Toffler and Don Tapscott. This phraseology has become so trite that the mere mention of it makes some people gag. This is unfortunate because talking about paradigms helps us understand the new economy. If you are the type that will gag when you hear about paradigms, you might want to jump ahead to the next section. But for those of you who want to learn about the rules that you must follow for success, stay right with it.

By now, most of us have heard about a major paradigm shift in the economy—the movement from an industrial-based society to an information-based society. We are beginning to change the way we act in response to the shift to the information economy. As the information economy matures, we see smaller shifts within the larger shifts. Wireless is one such smaller shift. But how did we get there?

Within the information age, there have been three waves of computing: backroom, desktop, and mobile computing. Backroom computing started more than 40 years ago. Large corporations and government organizations needed new ways to process the mass of information they had. They installed computers with names like Mark I, ENIAC, IBM 650, and UNIVAC I. From these humble beginnings arose the information economy. In the 1950s and 1960s, the computers solved business, scientific, mathematical, and engineering problems. Even backroom computing went through generations as we saw the movement from vacuum tubes to transistors to integrated circuits. The result of these movements was that the functionality of the $2.5 million UNIVAC I computer could be duplicated on a circuit board for less than $500. Looking at it another way, had the same technological improvements applied to cars, a Ferrari that sold for $28,000 in 1950 would have cost $5.60 in 1980, and 29 cents in 1994. So instead of looking for a parking space in Manhattan, London, or Toronto, to save money, you would just abandon your car!

Eventually these backroom computers were networked over the entire organization. The computer networks got bigger, and so did the supporting organization. Backroom computing was characterized by centralized processing and control. Data migrated to the computer center where it was converted, processed, and stored.

By 1971, another technological breakthrough was to crack this cycle and empower the users. A change in chip architecture produced a programmable one-chip processing unit, or microprocessor. In 1975, you could buy the first microprocessor-based computer, the MITS Altair 8800, for $395. From these modest beginnings came the Timex Sinclairs, Commodore VIC 20s, TRS 80s, Apple IIs, and their progeny. Annual shipments of microprocessors (which are used in everything from toasters, to cars, to personal computers) grew from a few thousand in 1971 to about 200 million worldwide in 1981. When introduced in 1974, the Intel 8080 microprocessor cost about $360. In 1984, it cost less than $3. The dramatic drop in price parallels the equally dramatic growth in personal computers. Correspondingly, there was a movement from the computer center to the desktop. Starting in 1981, desktop computing led to local area networks, the flattening of organizations, client/server computing, rightsizing, and other variations on this theme. The plummeting prices of microcomputers (personal computers) and microprocessors is changing the way people are doing their job.

People want to move around and still have access to all the information they need to do their job. This need led to the growth of the portable personal computer market. Now you can buy a notebook computer that possesses an incredible amount of processing power. The move to mobile computing is evidenced by the notebook computer market. The fastest-growing market area for computer technology is notebook computers. This growth is manifesting itself in a demand for mobile computing, the ability to use the network from any location. People want to take their notebooks with them and connect to the LAN.

Like the LAN itself, wireless LANs will change the way people work. You will no longer need cubicles and buildings to work in. Truthfully, we only came together to work because the information and tools we needed were centralized. Home offices now have microcomputers, modems, facsimile machines, and laser printers. Soon we will access the information from anywhere. Wireless LANs are an enabler in the work of the future. Individuals will use wireless communications to access people and information anytime, anywhere. The only question is, "How quickly will this happen?"

You already saw that researchers like Forrester and International Resource Development predict tremendous growth in the use of wireless LAN technologies, and those who exploit the shift early on gain a competitive advantage.

Wireless transmission technology for LAN applications is still in an introductory stage, with infrared, UHF, and microwave radio having attracted the most attention so far. The primary goal of this technology has been to substitute premises cabling, although an untethered workplace is now a growing motivation. This untethered approach is supported by the industry's trend toward smaller portable personal computers such as notebooks, laptops, and palmtops.

Voice and data traffic will eventually travel along wireless PBXs and wireless LANs. As a premises wiring substitute, cordless telephones, PBX adjuncts, campus links, and 5.7-Mbps LAN systems are already available in the marketplace. There are some technical hurdles for radio in supporting useful distances, data rates, and flexibility, but these currently available systems demonstrate the feasibility within a range of applications. It is interesting to note that even with significant regulatory restrictions, radio is being vigorously applied and marketed. This alone provides some glimpse of its potential market value.

Another wireless technology, line-of-sight infrared transmission, has been applied to telephone headsets, campus telecommunication, and now LAN equipment. So far, market response to the infrared Token Ring systems that are currently available has been limited. This equipment requires optical transceivers that must be aligned, creating situations that can sometimes detract from office aesthetics. Campus links for telephone and data communication systems have been available and successfully used for many years.

Proprietary radio systems using general (industrial, scientific, and Medical) and licensed frequency bands already are on the market. A close examination of the limitations of radio has led most analysts to conclude that it will not substantially eliminate customer-premises cabling, but will become mostly an extension option for cabling systems.

The regulatory situation provides a major impediment. Frequency allocations for premises radio are still not globally settled, and support of 10- and 16-Mbps LANs require very large allocations of the electromagnetic spectrum. Accommodating a 100-Mbps LAN will be significantly more difficult for both allocation and technical solution. Already, regulators and developers recognize the need for a wide block of the electromagnetic spectrum, and standard committees are working to establish radio LAN standards.

In spite of the demand for allocations and standards, the difficulties in establishing them will result in a delay of a couple of years for a standards-based LAN. Once established, standards-based equipment will improve the radio LAN market segment, just as it improved other segments of the computer marketplace. In this time frame, however, 10- or 16-Mbps radio LAN equipment will then be one of many extension options for the newer switching technologies.

Currently available wireless systems do not substantially eliminate cabling. They rely on cabling for system interconnection (for example, bridges), and they use the wireless to accommodate user portability and flexibility. Figure 1.5 provides an example of this.

So, wireless exists. Why would you want it? If you are reading this book, no doubt you have an interest in wireless, but you want to learn the pros and cons for the technology. Let's look at the current arguments for wireless communications.

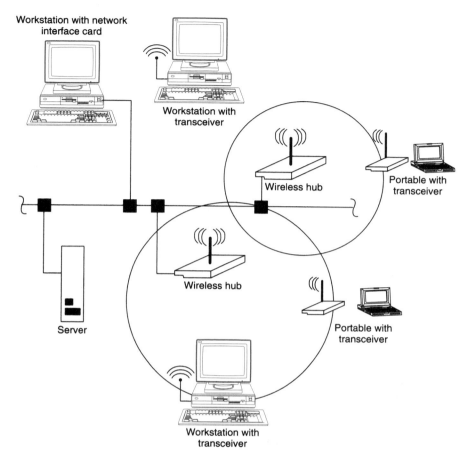

Workstation with network interface card

Workstation with transceiver

Wireless hub

Portable with transceiver

Wireless hub

Server

Portable with transceiver

Workstation with transceiver

Figure 1.5 Wired and wireless connections.

Why Wireless Communications?

Wouldn't it be nice to leave behind the physical constraints of your office or plant—your desktop personal computer, your electronic mail, your telephone—but never lose touch? Wireless LANs would enable us to roam the office and plant, yet still be in touch. It's an immensely appealing idea to a lot of people.

Wireless LANs are not a new idea. Early LANs were analog, broadband-based networks. *Broadband* means that the LAN cable splits into channels (think of cable television), and you tune in the channel you want. In 1984, consumers bought broadband LANs to connect personal computers. These

LANs failed because of their complexity and resulting unreliability. Today's LAN technology is baseband-based, which means the cabling media is one channel. It is this very cabling that resulted in a resurgence in wireless communications.

The resurgence of wireless communications began about 1987 with in-building, peripheral-sharing wireless devices. The products were easy to install but insufferably slow. Slow would be a kind word for some wireless network transmission rates, with early products chugging along at throughputs in the 19.2-Kbps to 1-Mbps range. The poor performance was a tangible factor in the relatively low sales of desktop wireless LANs. When comparing LAN wireless technologies, two crucial issues help define all wireless devices: data rate and range. Electromagnetic science dictates that the higher the data rate, the shorter the range. For example, some infrared devices that rely on a clear line of site for transmission have high data rates but extend short distances, somewhere in the range of 20 to 50 feet. Other infrared products use a diffused signal that floods an area but has a weaker signal. Radio technology has low data rates but can transmit 1 to 2 miles.

Whether a wire is made of copper or glass, transmitted signals share the space only with other data. Wireless transmissions, on the other hand, share space with sound, light, radio, and radar, among other things. How the data moves and the physical points it can reach are determined by the service supplier. The way the data shares its invisible wire with other transmissions is dictated by the Federal Communications Commission (FCC) in the United States and the Canadian Radio and Telecommunication Commission (CRTC) in Canada, which allocate rights to the electromagnetic spectrum (see appendix C).

The lower bandwidth of the electromagnetic spectrum carries audible sounds, while the upper reaches carry light and subatomic particles. Bandwidth is already assigned to local governments and to industries such as petroleum and transportation. Other parts of the spectrum are open but relatively crowded. For example, indoor security and paging systems also use the spread-spectrum radio (902–928-MHz) band. In the United States, the FCC plans to clear the 2.400–2.500 and 5.750–5.825-GHz space for wireless communications, but earlier settlers object to relocation because of the trouble and expense. Thus, industrial users of the spectrum are suddenly holding valuable pieces of electromagnetic property.

The battle for transmission bandwidth will be fought in the courts and legislatures. In the United States, the FCC has promised several segments, notably 930–931 MHz and 2–2.4 GHz, to companies willing to pioneer services at those frequencies.

Megahertz and gigahertz, all this is Greek to me. Well, don't worry. You'll learn about these terms and the different wireless technologies in chapter 3.

At this point, all you need to know is that it works, and the FCC, the CRTC, and others are striving to control the electromagnetic spectrum.

So wireless is technically viable and potentially legally possible, but why would you need it? Wireless is for use in locations that are difficult to wire, such as the factory floor, buildings with asbestos, and temporary installations such as Desert Storm, disaster recovery sites, and trade shows. Today, businesses have an increased demand for fast, accurate information. Well, yes, these are the positive reasons for wireless, but there must be negative reasons, you say. What about higher start-up costs?

Sure, wireless technologies have higher start-up costs. But a primary incentive for wireless LANs is the reduced cost of installing, moving, and changing networks. Network workstations are constantly on the move. The major cost incurred by all this movement is in the existing horizontal wiring. Once it's in place, it's there, whether the user moves away or not. Add to this the hidden costs of having the existing wire tested, new wire installed by a certified installer for a given building, possible remodeling costs, equipment storage costs, and the costs of downtime because wiring difficulties delayed the installation of a LAN. However, under any realistic analysis, ultimately these costs will outweigh the start-up costs of wireless technologies.

With every passing day, our society becomes more and more mobile. No doubt, some of you use pagers and public address (PA) systems to keep in constant contact with your employees. In many companies, the PA system constantly bombards employees with pages for phone calls or meeting announcements. If you are not used to it, the constant bombardment disturbs you at first, but you quickly adapt. How many of your corporations have embraced cellular telephone technology for employees? The advantages of the wireless data network are similar to those of the paging and wireless voice network: portability, mobility and economy.

Economically, there are advantages to the wireless data network, particularly communicating without the expense of land lines. Wireless networks eventually might access the existing telephone network, but users do not have to extend the wire network or conform to its physical constraints. Rather, airwaves become the transmission medium. As we will see in the next section, already there is wireless LAN technology for interoffice networks, reducing the wires between offices. There are even greater economic advantages when a radio frequency network is used for access by a salesperson in a car many miles from the office.

Concerns that influence your purchasing decision for wireless technologies are signal penetration (are there thick walls?), transmission rate (wireless probably will never match wirebound), interconnectivity (new emerging technologies), user transparency, and costs. Let's look at some of these technologies and some of their characteristics.

Wireless Data Technologies

Wireless networks have exploded onto the scene, and in the aftermath, they have generated a great deal of confusion about the different types of wireless data networks. Basically, there are two primary types of wireless communications: wireless local or campus area networks (WLAN) and wireless metropolitan and wide area networks (WWAN). These are not new distinctions. They currently exist in the wirebound world.

Third-party service providers characterize WWANs. These providers charge tolls based on network usage. The geographic coverage of these providers and their networks is vast, being measured in miles or kilometers. These networks might cover a city, province, state, country, or the planet. Because of the vast distances, speeds are relatively slow—about 20 Kbps. Most of us are familiar with WWAN examples: cellular telephone, paging systems, and packet radio systems. The products representing this segment parallel the wired wide area networks, exemplified by the regional Bell operating companies (RBOCs) or information services such as America Online, CompuServe, GEnie, and Prodigy.

On the other hand, private ownership and the absence of toll charges characterize WLANs. The geographic coverage of these providers and their networks is limited, being measured in feet or meters. These networks might cover a plant, an office, a building, or a campus. Because of the short distances, speeds are relatively fast—2 Mbps or faster. Wireless bridges and PBXs typify WLAN products.

While this book deals only with wireless communications for local and campus area networks (WLANs), we will briefly mention mobile data communications. These products help companies replace buildings with notebooks and communications. This trend is materializing in personal communication systems and wireless networks.

There are many wireless data technologies. Presently there are, among others, technologies based on:

- Spread-spectrum radio.
- Infrared communications.
- Mobile radio.
- Meteor burst communications.
- Microwave.
- Very Small Aperture Terminal (VSAT).
- Mobile satellite communications.
- FM squared.
- FM sideband.

Spread-spectrum radio

Spread-spectrum is one of the primary wireless LAN technologies and provides the basis for a great deal of this book. You can grasp the gist of this technology fairly easily. You can produce and send radio waves across a wide range of frequencies, starting as low as 10 kHz and as high as 50 GHz, as shown in Table 1.1. The radio waves produced for spread-spectrum typically are 902–928 MHz. Think of spread-spectrum transceivers as sending data just the way a radio station transmits music to your tuner. This analogy is helpful but only partially correct. The differences make spread-spectrum ideal for transmitting data, preventing one device from interfering with another and providing a measure of security.

In spread-spectrum, the transmitter distributes the data across a wide band of radio frequencies, hence the term spread-spectrum. The transmitter adds extra bits to the signal to create a spreading pattern. The receiver, using the same extra bits as a key, decodes the spreading pattern and obtains the original data.

By distributing the data across radio frequencies, spread-spectrum can provide acceptable data communication rates, usually 2 Mbps. Also, interference on one part of the frequency band will not affect another transmission.

Spread-spectrum also provides a measure of security. Only the intended device can decode the spreading pattern. Every other device hears noise. That's all you need to know about spread-spectrum for now. You'll learn more about spread-spectrum in chapter 3.

Spread-spectrum vendors include Caliber Tek, Cardinal Technologies, Cylink, Digital Ocean, Gambatte, Hewlett-Packard, Hillier Technologies,

TABLE 1.1 Radio Frequencies and Bandwidths

Frequency	Band
< 30 kHz	Very low frequency (VLF)
30–300 kHz	Low frequency (LF)
300 kHz–3 MHz	Medium frequency (MF)
3–30 MHz	High frequency (HF)
30–300 MHz	Very high frequency (VHF)
300 MHz – 3 GHz	Ultra high frequency (UHF)
3–30 GHz	Super high frequency (SHF)
> 30 GHz	Extremely high frequency (EHF)

InfraLAN Technologies, Laser Communications Inc., Metricom, Motorola, NCR, National Semiconductor, Nuvotech, O'Neill Communications Inc., Persoft, Pure Data, Photonics, Proxim, Scientific Technologies Inc., Smoke Signal Data Systems, Solectek Corporation, Spectrix, Symbol Technologies, Telesystems/Telxon, and Windata.

Infrared communications

Another of the primary wireless LAN technologies that forms much of the book is infrared communications. Customers have used it primarily for short-haul, local communications. Most likely, you have used infrared transmitters many times and not realized it. The remotes for your TV, stereo, and other appliances use infrared signaling. As you will attest, these devices are low powered (two "AA" 1.5-V batteries last several months). The devices are also highly directional, typically point-to-point. (Try pointing them at the ceiling.) In addition, they are one-way devices, and separate devices are needed for two-way communications.

Infrared radiation is the region of the electromagnetic spectrum between visible light and microwaves. Lasers and masers can emit highly monochromatic (single-wavelength) infrared radiation. These waves are invisible but, like visible light and radio waves, can pass through empty space. Therefore, you can heat molecules to create heat rays, which in turn can carry information. In electromagnetic science, there is a direct relationship between the length of the waves and the frequency. You'll learn more about this relationship in chapter 3. Table 1.2 shows the various lengths of the waves of the electromagnetic spectrum.

Infrared has some serious limitations. Because transmitters/receivers must be able to see each other to communicate, their connections are fragile. Anything, even a single sheet of paper, that blocks the beam between two nodes will immediately render that link useless. Light transmission also is subject to fading and absorption.

There are some positive benefits to infrared. For instance, infrared can transmit over greater distances than spread-spectrum. Infrared systems are easily portable, can be set up in less than an hour (usually), and they don't call for licenses or right-of-way permits. Again, you'll learn about infrared communications in chapter 3. Infrared vendors include Hewlett-Packard, InfraLAN Technologies, Laser Communications Inc., and Solectek Corporation.

Cellular telephone modems and adapters

Vendors are developing WWANs incorporating cellular telephony. Unlike other specialized options of wireless data transmission, using the cellular network allows you to access a high-speed modem from any of the locations where cellular service is available. Through the cellular network, you have

TABLE 1.2 Electromagnetic Wavelengths

Type of radiation	Typical wavelengths
Radio waves	328 feet (100 meters)
Television waves	1.65 feet (.5 meters)
Radar waves	.03 feet (.01 meters)
Microwaves	.003–0.328 feet (.001–0.1 meters)
Infrared waves	.000,164 feet (.000,05 meters)
Visible light waves	.000,001,5 feet (.000,000,5 meters)
Ultraviolet waves	.000,000,3 feet (.000,000,1 meters)
X-rays	.000,000,000,03 feet (.000,000,000,01 meters)
Gamma rays	.000,000,000,000,3 feet (.000,000,000,000,1 meters)
Cosmic rays	.000,000,000,000,03 feet (.000,000,000,000,01 meters)

the capability to connect to both private and public communication networks such as Tymnet, SprintNet, and CompuServe, which are dial-in networks. Common applications include remote LAN connectivity, remote e-mail, remote control of an office computer, connecting to online services such as America Online, CompuServe, Prodigy, and other electronic bulletin board services.

The allocation of frequencies for cellular are in the 825–845-MHz and 870–890-MHz bands. In each band, the channels use a 30-kHz separation, and 21 channels are allocated to control channels (of possible 666 duplex channels).

Some cellular equipment vendors are Cellabs, Cellular Data, Hayes, Magic Soft/WordPerfect, Microcom, Mitsubishi, NovAtel, Oki, Omnitel, Pacific Communication Sciences Inc., PowerTek Industries, SpectrumCellular, Telebit Corporation, Telular, Toshiba, U.S. Robotics/Touchbase, and Vital Communications.

Mobile radio

Several vendors offer mobile radio products that work on a dedicated radio channel, usually below 1 GHz. The data capacity, coverage, and reliability of these systems depend on the chosen frequency. You can transmit more data at higher frequencies, but you limit the coverage to line of sight from the transmitter. Depending on the chosen frequency, data rates range from 100 bps to 9.6 Kbps.

When using a dedicated radio channel, you can tailor your system to suit the intended application. You also "own" the signal distribution system and thus avoid the headache of having to repeatedly negotiate distribution rights with suppliers. On the downside, there is a relative scarcity of electromagnetic spectrum, particularly in urban areas, making it difficult to gain exclusive access to a radio channel. Mobile radio networks are WWANs and have found use in police and taxi applications.

Mobile radio equipment vendors include Astea International, Dataradio, E.F. Johnson, ElectrocomAutomation, Electronic System Technology, Ericsson GE Mobile Communications, Gandalf Technologies, Glenayre Electronics, Granite Communications, Hadron, HAL Communications, II Morrow, Intermec, Kustom Electronics, LXE, Laversab, Linx Data Terminals, Megadata, Monicor Electronics, Motorola, Multipoint Networks, NCR, Norand, Racotek, Telxon, and T.I.

Meteor burst communications

Meteor burst communications is a limited application for radio transmission systems. Using a predefined path, you transmit a radio wave up through the earth's atmosphere into space. Blocked data bursts from the transmitter and travels along these defined paths into the direct path of free space meteors. The data then bounces back to earth receivers. Applications using this technology tend toward telemetry, weather, and navigational systems. However, you could use any block mode data transmission in the meteor burst spectrum. Because of the limited applications for meteor burst communications, there are few vendors, but the primary vendor is MCC.

Microwave

Most of us are familiar with some form or another of microwaves. Microwaves have a number of applications, such as ovens, radar, and telecommunications. Because this book is about telecommunications, and not cooking or navigation, let's start there. Again, this is one of the primary wireless LAN technologies covered in this book.

Microwaves, so called because of their short wavelengths and high frequency, can carry information for networks. They, like other electromagnetic radiation, will travel, or propagate, through space. Table 1.3 provides typical distances you can transmit before you require microwave repeaters. In LAN applications, these microwaves must be directed or guided through space using an antenna. Because of their high frequency, microwaves can be accurately directed in a narrow beam from one transmitting antenna to a receiving antenna. These antennae perform the same function as the

**TABLE 1.3 Typical Microwave
Transmission Distances**

Frequency	Distance (approximate)
2–6 GHz	30 miles (48 kilometers)
10–12 GHz	20 miles (32 kilometers)
18 GHz	7 miles (11 kilometers)
23 GHz	5 miles (8 kilometers)

more common types of antennae but are different in construction. A receiving antenna similar to a transmitting antenna is used to receive microwave radiation, which is then directed to the appropriate device for further processing.

Nondirectional microwave technology overcomes infrared's line-of-sight problems and delivers a bandwidth in excess of either infrared or spread-spectrum. Another advantage of microwaves over ordinary radio waves is that microwaves, which have a higher frequency, can carry more information because information capacity is directly proportional to frequency. In chapter 3, you'll learn about microwave technology.

This technology is fairly specialized. Microwave equipment vendors include California Microwave, Digital Microwave Corporation, International Microwave, Laser Communication, Microwave Data Systems, and Microwave Networks.

Mobile satellite communications and Very Small Aperture Terminal (VSAT) technology

In the past, countries launched and operated satellites. Not so long ago this changed, and satellite services were used by large organizations with huge amounts of data, video, or voice traffic. Large antennae with complex electronics software and hardware were typical. The equipment was single-use and highly specialized. Today, that is changing.

Those days are past, due largely to recent advancements in earth station design. A major new advancement is the Very Small Aperture Terminal (VSAT) using spot-beam transmission, which is applicable to wireless data applications, where the size and cost of the earth station has been greatly reduced.

Satellite communication is remarkably easy to understand. Visualize a triangle (see Figure 1.6). At the top of the triangle is the satellite, and the lower two points are the earth stations. In its simplest form, in a point-to-

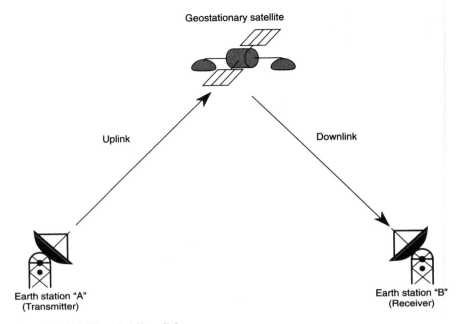

Figure 1.6 Satellite uplink/downlink.

point configuration, one earth station is the transmitting location (the uplink) and the other is the receiving location (the downlink). To create a
more typical configuration, just add more downlinks, all receiving signals
from an uplink. Furthermore, each downlink earth station is independent of
all others for installation, operation and maintenance, and can be moved
without affecting the remaining sites. You access satellites by transmit-and-
receive earth stations. Transmitters and receivers can be commanded to
switch on or off, or to tune to a specified satellite channel. Therefore, you
could install a receiver on a truck to transmit and receive current information about location, fuel consumption, and vehicle fitness.

Today, satellite service providers plan, design, and offer services to anyone within the coverage area of a satellite. In some cases, the antenna coverage is deliberately designed to cover a very wide area, while in other
cases, the use of a normal coverage pattern is sufficient to reach a sub-
stantial and common market. You can find examples of this in Europe,
North America, and in the various regional systems implemented around
the world.

Satellite technology offers several varieties of channel for data transmission. They range from full-transponder (services such as FM2, FM3, SCPC)
to low-powered 3.5-kHz channels above video, so called audio subcarriers
(FM/SbC). There also is a selection of satellite frequencies, namely C band,

at 6/4 GHz, and Ku band, at 14/12 GHz (uplink and downlink respectively). Table 1.4 shows the range of bands assigned by the Federal Communications Commission (FCC) for satellite communications.

Satellite technology offers the following benefits:

- Improved cost-effectiveness because service rates are distance insensitive and spread over the entire network of receiving locations.
- Consistent quality of communications because of fewer hardware components in the overall network.
- The network can expand easily within the satellite's footprint.
- Smaller VSAT earth stations that are less costly and easier to install.

Vendors are American Mobile Satellite Corporation, AT&T Tridom, GTE Spacenet, Hughes Network Systems, Inmarsat, Mobile Datacomm Corporation, Qualcomm, and Scientific Atlanta.

FM squared

Arguably, the most advantageous data broadcasting method today is FM squared, or FM over FM. FM squared is a signal modulation technology. Instead of placing the subcarriers above the FM stereo band (FM sideband),

TABLE 1.4 Satellite Communication Bands

Frequencies (up/downlink)	Actual range (GHz)	Band	Use
6/4	Uplink: 5.3925–6.425 Downlink: 3.7–4.2	C	Commercial
8/7	Uplink: 7.9–8.4 Downlink: 7.9–8.4	X	Military
14/11	Uplink: 14.0–14.5 Downlink: 11/7–12.2	K_u	Commercial
17/12	Uplink: 17.3–17.8 Downlink: 12.2–12.7	K_t	Future
30/20	Uplink: 27.5–30.5 Downlink: 17.7–21.2	K_a	Commercial
30/20	Uplink: 30.0–31.0 Downlink: 20.2–21.2	K_a	Military
44/20	Uplink: 43.5–45.5 Downlink: 20.2–21.2	Q	Military

the signal is loaded together and modulated. Like FM sideband, many applications abound. For instance, NOVANET uses FM2 to reach Canadian Press' 500 locations, carrying news stories.

FM sideband

The original monaural FM radio system occupied only 15 kHz of the FM baseband. Thus you could easily add a stereo channel to bring the total baseband up to 53 kHz. In the 1950s, U.S. service providers began to add an additional portion to the FM baseband to distribute Muzak and other background music.

The following are frequency bands containing one or more subcarriers that are outside the normal tuning range or response of consumer FM receivers: in the United States, a subsidiary communications authorization (SCA) channel; in Europe, a radio data system (RDS); and in Canada, a subsidiary communication multiplex operations (SCMO) channel. SCA, RDS, and SCMO are these additional portions of the FM baseband. And you can transmit a variety of signals, including speech, music, tones, and data over these channels. These channels are a frequency band in the baseband of the FM signal containing one or more subcarriers, each having a frequency in the range of 53–99 kHz. The subcarrier in North America has a frequency of 67 kHz and 92 kHz, while in Europe the frequency is 57 kHz. Transmission rates of 19.2 Kbps are possible with advanced equipment.

This technology is primarily used for data broadcasting, the transmission of information over a wide area to many different locations. You might want to broadcast "hot" credit card numbers to retailers or new interest rates to bank branches.

Some equipment vendors are Cue Networks, Sherbank Marketing Services, and Silent Radio International. As mentioned, you'll explore the different wireless local area networking technologies in chapter 3.

Wireless Local Area Network Applications

There are three primary applications for WLAN technology. There is the wireless desktop application, the wireless building-to-building short-haul connection application, and the portable mobile connectivity application. There are many examples of the first two applications today, while portable local mobile applications are just developing.

Within WLAN applications, already there are various applications with different data rate requirements, cost and size considerations, and mobility needs. For example, a hand-held data communicator for inventory control must be lightweight, small in size, mobile, and able to handle data rates of only several hundred bits per second. Contrast this to a wireless LAN connection requiring several million bits per second, where weight, size, and mobility are not as important.

Wireless desktop applications extend wirebound LAN applications. Basically, these applications integrate wireless desktops with wired LANs. Products must provide high speeds (at least matching wired LAN speeds) for short distances. The primary players are Motorola's Altair, NCR's WaveLAN, and Windata's FreePort products.

Wireless building-to-building connection applications provide LAN-to-LAN interconnections between buildings and can replace leased land lines and T-1 links. Products must provide fairly high data rates for longer distances (up to a mile). Again, the primary players are Laser Communication's LACE, Microwave Networks' MicroNet, and Persoft's Intersect.

As revealed, the portable local mobile applications are just evolving. The most common application is inventory control, using hand-held communicators with scanners. Products must be lightweight, small in size, mobile, and able to handle data rates of only several hundred bits per second. Data rates will remain low for the foreseeable future because of the direct relationship between data rate and power consumption. This application will soon take off because of the proliferation of portable computers in the office environment. Users will want to connect their notebooks, laptops, and palmtops to the network when they are in the building. The primary players are Photonic's Photolink, Proxim's RangeLAN, and AIRONET's ARLAN.

In chapter 5, you'll look at wireless for network replacement and enhancement. In addition, you'll learn about the applications in Part 3 (chapters 6 and 7). As well, you can find product descriptions in Appendix A.

Future Trends and Issues

Later in this book, you'll also examine future trends in wireless LANs. Some future trends and issues you look at are standards, health and safety, and security.

What about standards?

As already suggested, standards are necessary for the wireless LAN market to really take off. Without standards, you have to rely on proprietary software and hardware. Proprietary products usually translate into higher prices. Hopefully, the developing standards will provide interoperability of wireless network products. Once consumers, vendors, and governments agree on standards, it should mean that compliant products will be available. Standards tend to promote interoperability and drive down prices. The discussion of standards in chapter 4 will help you understand the issues, progress, and the direction of evolving standards in the wireless local area networking marketplace.

Health and safety

Another consideration for wireless LANs is health and safety. Concerns have been raised over the years about the danger of exposure to low-level electromagnetic fields. The past five years have seen a revolution in public awareness of electromagnetic fields. In the past, computer equipment had to meet strict guidelines set by governments for shielding users from electromagnetic radiation.

Scientists have not established an ironclad link between exposure to electromagnetic fields and ailments such as cancer, but there is enough circumstantial evidence to worry many people. In chapter 8, you'll explore these issues and more from a health and safety viewpoint.

Security

Another major issue cited by many potential purchasers of wireless technology is security. There is a widespread belief that wireless LANs are insecure. Generally, unencrypted, wireless LAN transmissions are easier to eavesdrop than conventional networks. Therefore, it might be necessary to encrypt them, although some vendors argue that it is easier to intercept the electromagnetic fields around cables than it is to pick up some microwaves. On top of that, they argue, the wireless transmission is encoded. If you intercept the signal, you still need the encoding scheme to understand the data. Also, wireless vendors and others argue that because there is an abundance of wired LANs, there is a corresponding abundance of conventional network monitoring equipment, which makes interception of data on wired LANs easy. For instance, you can get a packet analyzer for Ethernet from most bulletin board systems.

Further, wireless vendors argue their networks are more reliable than cabled networks. Any administrator will tell you that the majority of networking problems are traced to wiring problems. And other threats exist. Radio transmitters presently are expensive, making them a target of thieves. Message interception, system unavailability, and theft of equipment are just a few threats to your wireless network. In chapter 9, you'll discover security exposures for wireless LANs and probe some possible controls.

We called this part of the book the concepts. This is the end of the background material, including a survey of wireless data communications, the issues, and the potential market for it. In the next chapter, you'll review the components of a local area network (wired or wireless), and you'll discover those components that can be replaced to build a wireless LAN.

2

Introduction to Local Area Networks

The use of local area networks in business organizations has seen incredible growth since 1988. LANs can be found in almost every size and type of organization, and they provide the computing power for everything from occasional printer sharing to running business-crucial operational and financial systems.

At least part of the reason for this growth is that the task of interconnecting computers has become far easier. In the past, installing a network and getting it to work successfully required two parts good luck and one part magic. Today, network interfaces are much easier to install and configure, (that is, to set up so that they function correctly with their host computer and its software). Network interfaces also are more cooperative about "talking" to each other, even between hardware from different manufacturers. As a result, setting up a LAN is now a largely straightforward task (although a little good luck still helps).

Lower network costs also have contributed to LAN growth. The previous high cost of network components limited the availability of connections to those with very specific business needs (for example, access to an expensive disk storage device or high-speed printer). However, the price of network components has dropped dramatically, while the number and level of services they enable has greatly increased. This makes it far more affordable, and therefore, more cost-justifiable, to connect most or all of the personal computers within a department or an office. Once connected, users can take advantage of simple functions like sharing disk space or laser printers. Connection to the LAN also allows them to share files, exchange electronic mail, arrange scheduling, and collaborate on group projects.

But one of the biggest reasons for the growth of LANs is that they can make good business sense. A LAN provides impressive levels of computing capacity at lower costs and with greater flexibility, when compared to traditional mainframe computing platforms. A centralized mainframe computer is typically very expensive, both for initial acquisition and for capacity upgrades. And upgrades usually come in large, expensive chunks. That means you hold off on installing the upgrade until you absolutely need it, then you probably get more capacity than you need. On the other hand, personal computers and the related network hardware are very cost-effective. You can build processing power and storage capacity rivaling many mainframes. You can add more file servers, increase disk capacity, add more printers, or just reconfigure the whole network more easily and cheaply than changing a mainframe. In many circumstances LANs, really do provide the most "bang for the buck."

Despite how common LANs have become, what it is they do is often mysterious to the average user. Obfuscatory verbiage abounds, along with acronyms, abbreviations, and an endless collection of standards with names that look suspiciously like extracts from the penal code (". . you have been charged under Section 802.3 with having a promiscuous Ethernet connection . . ."). All this makes it difficult to understand what a wireless LAN can do and to evaluate if and when one can be of use to your organization.

So to better set the stage for your study of wireless LANs, in this chapter you'll explore key network concepts and the major components of any LAN. You'll review network concepts, including servers, clients, and the interfaces and media that tie them together. Then, in the next chapter, you'll see how wireless components are used as part of the LAN.

What Is a Local Area Network?

The term network refers to interactions taking place between defined units. For our purposes, a network refers to a collection of microcomputers linked by some type of medium. The link can be for one specific purpose, such as to share large disks or fast printers. Invariably, once the lines of communication are opened, other uses for the network appear. When computers can communicate, they can perform many useful tasks such as exchanging electronic mail for their users, accessing host computers through a network gateway, or even competing in games and simulations with other network users. But the key feature of any data network is the interconnection of two or more computers.

The connection between computers requires a medium that is used to transport data. The most common way to connect computers is through the use of electrical wire or optical cables which are known as *bounded media*. The term "bounded" is used since, for all practical purposes, the energy used to exchange data remains within the medium itself. (Most bounded media do permit energy leakage to a greater or lesser extent, but this is not key to their proper functioning).

Bounded media such as cables contrast with *unbounded media* like radio frequency links, microwaves, and infrared technologies. Unbounded media depend on the radiation of energy in order to exchange data. Devices using an unbounded medium are design to transmit and receive data without the use of cables, in the same way a radio transceiver (like a "walkie-talkie" or "handi-talkie") is used to exchange voice without the use of wires. For the most part, the choice of bounded or unbounded media is merely an implementation decision. The choice specifies how computers will be connected, not what someone actually does with the connection.

The term "local area network" also can be defined in many ways, each one subject to an amount of pseudoreligious fervor. Our meaning of LAN is a network whose components are within the complete control of your own organization. We also focus on LANs made up mainly of personal computers and their related hardware, although our definition of a LAN can include any type of computer resource. We also include campus area networks (CANs) involving links between buildings at one company location. Our discussion of LANs in general, and wireless LANs in particular, will be within this context.

This definition is designed to exclude networks made up of segments that are under the control of others, such as a data communication line or satellite link rented from a third party in order to connect offices in Toronto and San Francisco. These are usually thought of as wide area networks (WANs), and while they might involve wireless technology, such links are outside the scope of this book. Similarly, links within a city (known as metropolitan area networks or MANs) involving third-party carriers such as your local telephone company are not included in our definition of a LAN.

Any LAN consists of the following major components:

- Network nodes (also called endpoints).
- A connecting medium.
- Various levels of network communication protocols.
- (Optionally) internetworking devices (which are really just a special type of node).

You'll learn more about each of these components in the following sections.

Local Area Network Nodes

A node on a LAN is simply a device attached to the network at a given point. For wired LANs, the attachment is a physical connection between the network and the device, while for a wireless LAN, attachment can refer to a radio or infrared link. In either case, the device communicates through an interface to the chosen medium, whether that medium is bounded or unbounded.

The attached device is normally a computer, but it can also be an internetworking device used as a go-between for different portions of a network or between different types of networks. You'll see more about internetworking devices later in this chapter.

The network interface card

Between the computer and the connecting medium sits the network interface card (NIC). The NIC serves as the gateway between the computer's internal components and the network itself. One side of the NIC connects to the computer's internal bus, giving it access to the memory and the processor. The other side acts as the interface to the network-connecting medium.

The NIC is faced with two sets of rules for polite conversation. One set must be followed by any device accessing the computer's internal bus. The other set of rules is required of any node wanting to make use of the network. The rules to be followed are well defined and specific to the type of network the NIC interacts with, as well as the type of computer where it is installed.

Within these constraints, the NIC responds to requests from software to send messages across the network. The software provides data, which the NIC formats according to the rules for transportation across the network. For example, the NIC includes the required network addresses and related information. Then, when the network rules permit, the NIC sends the message.

The NIC also listens on the network for messages intended for its reception. It captures the message, checks the contents for compliance with the rules, then passes it on to its controlling software.

Network roles

Computers attached to the network usually fill one of two main roles. They function either as a:

- Server computer, providing services to the network such as storing files, printing data, or connecting to host computer or other network.
- Client computer, using the available services from the network.

Some data networks emphasize clear and distinct relationships between nodes. In these cases, referred to as client/server networks, a node is either a server or a client, but not both. A server node, usually a high-capacity, high-performance computer, will offer services such as making its disk drive available to store files on behalf of client computers. The client nodes, as you might expect, eagerly take advantage of the server's offerings. Figure 2.1 provides a depiction of a client/server network.

PC Client

File Server

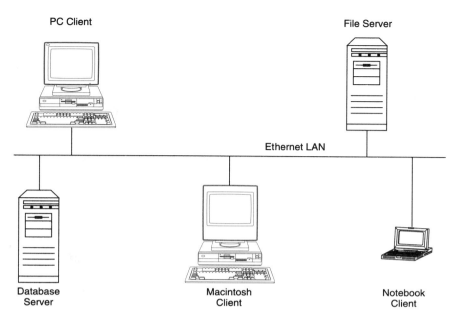

Ethernet LAN

Database
Server

Macintosh
Client

Notebook
Client

Figure 2.1 Example of a client/server network.

However, on other networks, a node can act as either a client or as a server, or even both at the same time. These are known as *peer-to-peer* networks. On a peer-to-peer network, each node has equal status and importance. Each computer can offer access to any resources connected to it (for example, disks or printers) as well as access other nodes offering their services. Peer-to-peer networks are often used for informal sharing arrangements (for example, between members of a department or work group). Figure 2.2 shows a model of a peer-to-peer network.

Services offered by nodes

Offering network services involves accepting requests from client computers on the network, performing some type of processing to service the request, and delivering the results back to the client across the network. A program on the server listens for network requests and knows how to deal with them. It then makes use of the network to return the results to the client.

A benefit of using distinct clients and servers is that you can separate some type of computing tasks, like file storage and retrieval, from other tasks such as calculating formulas in a spreadsheet. Using separate computers for separate activities allows you to optimize each one's performance for the task at hand. For example, a file server can be configured so that it

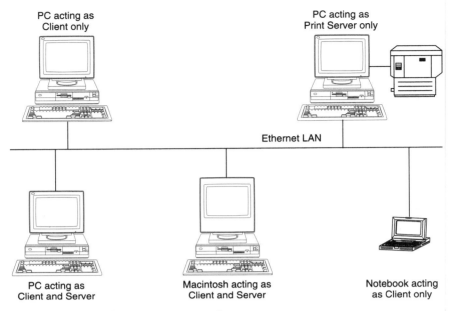

Figure 2.2 Example of a peer-to-peer network.

stores and retrieves files as quickly as possible, without having to be good at performing other general computing tasks like compiling programs.

Server computers can provide many different types of network services. For example, general-purpose computers are computers on a network that allow users to access them in order to perform a variety of tasks. Tasks include logging on, executing specific commands or procedures, transferring files to and from another computer, or sending and receiving electronic mail. Many types of computers using operating systems with names like MVS, VM, OS/400, VMS, and UNIX provide general services to the network.

Aside from the general-purpose computer, there are also many special-purpose servers. These computers are designed to perform one or two main tasks very efficiently and effectively. Some examples of servers include:

- File servers.
- Print servers.
- Database servers.
- Other specialty servers.

File servers. These are probably the most common type of servers on a LAN. Their purpose is to allow clients to store data on the file server, which normally has a high-capacity disk drive (such as from 1 to 10 Gigabytes of storage). To store and retrieve files, the client computer establishes a connection with the file server across the network. Once you establish the connection, programs running on the client computer can access the server's disk storage device as if it were a local disk drive physically connected to the client computer. Software running on the client computer evaluates any requests for file access and determines which ones must be sent across the network to the file server. The file server responds to client requests for information by supplying the data as requested.

Print servers. These servers are responsible for receiving requests from clients to receive documents or other types of files, and printing them on an attached device such as a high-speed dot matrix or laser printer. To produce printed output, a client computer redirects data normally intended for a local device and sends it across the network to the server. The print server maintains a waiting line (known as a queue) of documents and files for printing. It receives the data sent to it from the client, adds it to the end of the queue, and eventually prints the output.

Database servers. A database server is similar to a file server in that it services requests for data. However, while a file server is geared towards providing entire files, a database server is designed to provide specific portions of a file. For example, instead of delivering an entire payroll database file, a database server can select specific records, representing only those employees who match certain criteria. A client computer sends the server a request for information in the form of a database query. The database server is designed to very quickly locate the desired data, extract it from the database, and return it to the client node.

Other specialty servers. The flexibility of a network means that servers can easily be added not only to supplement existing offerings, but also to provide new services. Just about any computer can "hang out its shingle" and provide a number of services to the network. Servers have been built to provide gateways to remote host computers, to send and receive electronic mail, to perform searches for information across a worldwide network of computers, and even to authenticate users before they are allowed access to other network services. What makes a server a server is that it runs service software. Hence, a print server is a device, such as a personal computer, microcomputer or minicomputer, with print service software installed. Similarly, a database server is a device with database service software installed. So when a system administrator wants to create a new server, he or she simply installs the service software on a microcomputer, minicomputer, or mainframe computer and makes the service available to the network. Servers offer services to client nodes.

Client nodes

Each client node consists of a computer with the appropriate software that can make use of network services offered by server nodes. Software running on the client computer must be able to intercept and interpret local requests for remote services, formulate the appropriate message, and send the requests across the network to the correct server node. Once sent, the client node waits for a response. When received, the details are provided to the program or user making the request.

Almost any type of computer can be a client. This includes the general-purpose computers mentioned as being servers, plus a full range of mini and microcomputers.

Connecting Media

You now know about network nodes and the types of roles they play on the network. Now it's time to review how the nodes are connected together using connecting media. The connecting medium, which might be electrical wires, fiber-optic cable, or wireless, carries the data between nodes on the network.

Network topologies

Nodes can be connected in a variety of physical configurations. The layout of the connecting media between nodes on the network is referred to as the *network topology*.

In the simplest case, two nodes might have a link running directly between them. This is referred to as a point-to-point connection and is useful when only two devices need to exchange information (for example, a notebook computer copying files to and from a desktop computer).

However, the more general network case involves two or more devices sharing the connecting medium. Every node on the network is capable of receiving data from the connecting medium and of placing data on the medium for delivery to another node. This occurs, for example, when all the personal computers (PCs) in a department are connected to one common cable.

There are many ways in which nodes can be connected to form a network. Each of the common methods is known by a geometric shape such as a bus, ring, star, tree, or mesh. These shapes are far more easy to visualize when thinking about connections using bounded media such as electrical cable. However, they also are key to the way nodes on a wireless LAN communicate, as you'll see in chapter 3. Each configuration is described in the following sections.

Bus. If you are familiar with the internal path for data used within your personal computer, you already have experience with a bus. For a network, the bus is formed by connecting each node to a single medium. Each node can see all activity on the connecting medium. For physical connections to a wire cable, a T-shaped connector is used. This allows the electrical path along the cable to continue uninterrupted while the node gains access to the cable through the T-connector. In addition, the bus must be terminated in order to maintain the correct electrical characteristics. Figure 2.3 shows nodes connected in bus arrangement.

Ring. A network ring involves connections between nodes in the form of a circle. In order to connect to the medium, each node is inserted into the ring. As a result, all data passes through each node, which then sends it along the rest of the ring. The data therefore passes through every station during the course of making a complete circuit on the ring. Figure 2.4 is an illustration of nodes connected by a ring.

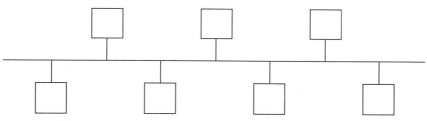

Figure 2.3 A bus network topology.

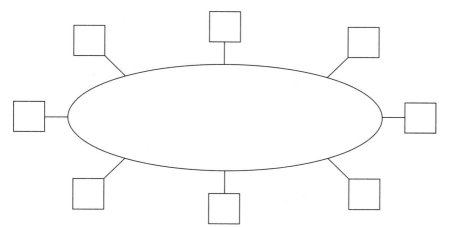

Figure 2.4 A ring network topology.

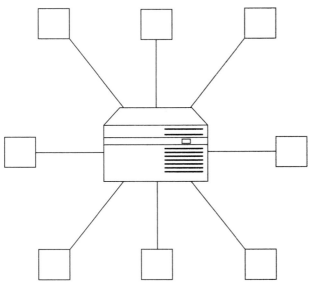

Figure 2.5 A star network topology.

Star. A network star is formed by connecting all nodes together through a central point, which is known as a switch, hub or concentrator. All data sent between nodes must go through the central point. The resulting data path therefore resembles a star, with a line from each node merging at a central core. Figure 2.5 illustrates the star topology.

Tree. A network tree is a special case of bus because each connection point can be split into other connection points. The best example of this configuration is a cable television system. The signal starts from the antenna, and fans out over the service area through successive taps.

Mesh. Because each node has its own point-to-point connection to every other node, the network mesh provides the most robust form of communication between computers. It's robust because if any one connection fails, then data can be routed through other nodes. In most cases, a full mesh is not implemented due to the expense involved in connecting each node to every other node. However, partial meshes also can be formed by providing some redundant data paths between nodes.

Controlling access to the medium

Sharing access to a common medium requires that rules be put in place. Whether the participants are two friends at a restaurant, eight guests at a dinner party, or hundreds of people in a town meeting, only one person can speak effectively at one time, while everyone else listens. Without these rules for the discussion, messages and meaning will be lost.

Computer networks are very similar. Whenever two or more nodes share one physical medium, there must be rules governing which node can transmit while the others listen. Every node must be prepared to follow these rules of conversation, otherwise data might be lost or distorted. And lost data means messages must be repeated (the computer equivalent of shouting louder), clogging the network and reducing efficiency.

The rules for use of a computer network revolve around when a node is allowed to access the shared connecting medium so that it can send a message on the network. Two main approaches of media access control are used, and these are known as contention and deterministic. Each is described in more detail in the following sections.

Contention. There are two main contention approaches to media access control. Each is used to ascertain if more than one node tries to use the network at the same time. They are both based on sensing the signal each node puts on the cable, earning the name of Carrier Sense Multiple Access (CSMA). CSMA can be used in one of two ways, known as collision detection and collision avoidance.

With collision detection (CSMA/CD), every node with data to send waits until no other stations are heard on the network. When the node detects silence, it transmits data. But it also listens to hear whether any other nodes are transmitting, in order to detect a collision. If it hears another node (determined by looking at changes in the voltage level) the NIC assumes that a collision occurred and that the data it sent was lost. The node then waits a random amount of time, and then attempts the transmission again. Because there is another node attempting to transmit, which has also detected a collision, the random waiting time generated by each node serves to prevent them from trying again at exactly the same instant.

The collision avoidance (CSMA/CA) approach is similar in that each node also waits for silence on the network before transmitting. However, it does not check for collisions by listening on the network for others transmitting. Instead, it evaluates whether the data made it through by waiting for a formal acknowledgment from the intended recipient. If an acknowledgment is not received within a certain period of time, the node assumes that a collision has occurred and that the data is lost. It then waits for silence, and after a randomly generated period of time, it tries again.

Contention methods are also known as *probabilistic techniques*. A node's access to the shared medium depends on the level of activity on the network, how many other nodes are also requesting access at the same time, and even the random time periods generated when collisions occur. You can therefore only determine how long you will have to wait for access

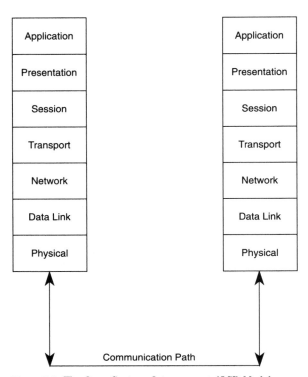

Figure 2.6 The Open Systems Interconnect (OSI) Model.

within a certain probability. This differs from the deterministic techniques described in the next section.

Deterministic. These approaches to media access control permit a level of certainty as to the maximum time a node will have to wait to get access to the network. They involve systematically passing a token between nodes. Only when the node has the token is it allowed to originate a message. The message goes across the network to the intended recipient, who acknowledges receiving the message to the sender. Once the acknowledgment is received, the sender must surrender the token to another node, even if it has more data to send.

Network Communication Protocols

The network nodes and connecting media form the physical portions of a local area network. Like a network of telephones and wires, or two-way radios and radio waves, we now have an infrastructure of computers that can exchange information. But what do they say to each other?

The answer lies in more rules, known as network communications protocols. People can use many different languages and protocols to communicate. Unfortunately, unless two people can speak the same language, they will have a great deal of difficulty in communicating. The same applies to computers. If they don't use the same language and protocols for network communications, then they cannot exchange data. There are many different network communication protocols available for use, and like human languages, it's not always possible to translate directly between them.

In an attempt to bring a level of standardization between protocols, the International Standards Organization (ISO) developed the Open Systems Interconnect Model (OSI Model). Figure 2.6 shows the layers in the OSI Model.

The model is intended to illustrate the separable functions needed for data communication among computers. The main benefit of the model is that it helps people to understand the data communication process in terms of functions operating at separate layers. Each layer plays a role in getting the data between the source and destination nodes. The OSI Model involves seven separate layers, but not many network communication protocols comply with this view. Instead, each protocol has its own view on the number of layers and the function of each.

Viewing a network in layers helps in understanding what happens when two computers communicate. For our purposes, following the OSI Model, or any other model for that matter, isn't important. Dividing the network communication process into manageable and understandable layers is. So let's use a layered approach to look at the network communication protocols in use when a client computer requests data from a file server:

- While using your computer, you start a spreadsheet program so that you can work with a file stored on a file server. After starting the program, you tell it to open the file. You don't care how the program gets the file. You just want it, so you let the spreadsheet program take care of the details. You pass a request to the layer below you (the spreadsheet program) and let it carry on.

- The spreadsheet program doesn't really know how to access the file server across the network. All it does is pass a request to open the file to the computer's operating system and let it work out how to get it. The spreadsheet program passes the request to the layer below it (the operating system) and lets it carry on.

- The computer's operating system, in cooperation with a network program loaded when you first started the computer, determines that the requested file is on a server across the network. It creates a request for the file and wants to send it to the file server node. But the operating system still doesn't know how to formulate requests to use the network itself. So it must pass the request on to the next lower layer, a program that knows how to send information across the network.

- The next lower layer is responsible for receiving the request for information, and addressing it to the right computer on the network. It places the file request in a packet of information and addresses the packet with the network address corresponding to the file server node. But this layer still does not have the capability to use the physical network. So it must pass the packet down to the next layer, which knows how to access the network to send and receive data.

- Now, down at the lower levels of the network model, you find a layer that receives the packet and can place it on the network connecting medium. This layer knows about issues such as media access control and the addresses of other physical nodes. But it still must rely on network hardware at the lowest layer to actually get the message through. So this layer passes the request down one more time.

- At the hardware level, a combination of programs and electronic components actually knows how to transmit to and receive from the network connecting medium. It takes the data from the layer above and sends it across the network.

At the receiving node, the same kind of process is repeated, only in the opposite direction. Each layer receives information from the one below, performs its task, and passes the details up one more layer. After the request is serviced, passed back down through the layers, across the network, and back up to you, you find that your spreadsheet is open and ready for use. The key concepts here are that each layer performs one set of functions, and each layer relies on the layers above and below it to exchange data.

As mentioned earlier, there are many different network communication protocols, and many individual components within each. They are fertile grounds for acronyms (for example, IPX/SPX with VLMs using NCP; TCP/IP with FTP, NFS, and UDP; etc. etc.). These can be the source of great confusion. However, when all is said and done, the role of all network communication protocols is to permit client and server computer nodes to reliably exchange useful information across the network.

Internetworking Devices

A simple LAN consists of a number of network nodes, a connecting medium, and a network communication protocol used to exchange data between nodes. But it is also possible to combine small, simple LANs into large, complex networks. Larger networks can help companies get a better return out of their investments in applications such as electronic mail and corporate databases. To do this, we use various types of internetworking devices.

Interconnecting devices pass messages from one LAN to another. They offer different levels of functionality and sophistication. For example:

- Repeaters, which simply extend the range of one LAN. A repeater rebuilds all the signals it hears on one LAN segment and passes them on to the other.

- Bridges, which can extend the range of a LAN or partition one busy LAN into two separate, more efficient sections. The bridge can examine hardware node addresses found in packets of information on each of the LAN segments where it is connected. Based on the addresses found, it will forward only those packets that need to be sent on to the other segment (unlike the repeater, which passes everything it hears, not unlike the office gossip). The bridge uses the data link layer of the OSI model.

- Routers, which connect LANs of different hardware types. A router, operating at the network layer, examines network addresses, which are conceptually separate from the lower-level hardware addresses examine by a bridge. Based on the network address-routing tables it maintains, a router can selectively forward packets on to another LAN.

- Gateways, which connect different network architectures by performing a conversion at the application level. A gateway maps from an application, say cc:Mail, on one computer to an application that is similar in function but differs in detail on another computer, say MHS in NetWare. The gateway itself uses all seven layers of the OSI model, plus all the layers of any proprietary architectures being connected.

In addition, the functions formerly associated with one device are now being combined into hybrid units. Devices with names such as brouters allow both bridging and routing functions to take place on a selectable basis, depending on the protocol and addresses involved.

How Does Wireless Fit In?

Now that you understand the basic structure of any LAN, you might wonder how wireless LANs differ from their wired counterparts. In the following chapters, you'll learn many details on how wireless LANs are built, how they operate, and how they can be used. As a preview, let's briefly look at some of the network components where wireless LANs might be different.

- Network nodes. As far as the host computer is concerned, the network interface card for a wireless LAN will appear similar to any other NIC. Of course, the NIC itself will be very different in that it will be designed to communicate over a wireless connecting medium such as radio waves or infrared.

- Connecting media. The most visible, or perhaps invisible, difference for a wireless LAN is that there are no copper wire or fiber-optic cables used to connect network nodes. Data is sent "across the ether" using one of the techniques you'll learn about in the next chapter.

- Network communication protocols. The protocols that are most closely tied to the physical components of the network are different for wireless LANs. However, for the most part, the higher-level network protocols used on a wireless LAN are similar or identical to those on a wired LAN. There might be some changes to account for a different type of physical link. For example, to deal with the generally slower speeds of a wireless LAN, there might be an adjustment in the amount of time allowed for a response from a server.

- Internetworking devices. As you'll soon see, wireless LANs are usually tied to wired LANs. Therefore, a connection is made through an internetworking device like a bridge, which is therefore part wireless, part wired. Wireless technology might also be used to form the internetwork link between two wired LAN segments. For example, you can establish an infrared link between LANs in two separate buildings on a campus. So internetworking devices can actually be a key component to using wireless LANs.

In this chapter, you learned the fundamentals of local area networking. Already, you can start thinking about where and when you might insert wireless devices into your network. In the next chapter, you'll focus on specific details of how wireless technology is used in local area networking.

3

Wireless Technology—Its Application to LANs

To better understand how you can use wireless LANs in your business, it helps to have an appreciation of how they work. That doesn't mean you need to be an electrical engineer in order to set up wireless LANs. But a little background information can go a long way in heading off misunderstandings and misconceptions. And some of the ideas might even seem familiar. Wireless communication works on the same set of principles, whether it's computer nodes on a LAN or the common portable telephone.

In this chapter, you'll learn some of the basic concepts of wireless communication, including the components needed, the transmission media used, and how the data is actually sent. You'll then discover how these concepts are applied to build different types of wireless networks.

Basics of Wireless Communication

Communication involves the exchange of information between two or more parties. Each unit of communication requires one party to originate or supply the information, some means to convey it, and another person to receive it. For example, in a conversation between two people in a room, one person forms words with their mouth. This action produces sound waves, which are simply vibrations causing changes in air pressure. Air carries the sound waves to a second person. That person detects the pressure changes with their ears, allowing them to hear the words. (Whether or not the words spoken and heard are meaningful depends on factors such as a common language, distractions like noise, and perhaps most importantly, clarity of thought.) Figure 3.1 illustrates this process.

But changes in air pressure is only one way to convey sound waves and the information they contain. The early telephone provides an example. Here sound waves are first converted to electrical signals by using a microphone. These signals are sent through a simple wire cable to the receiving end, where the signals go to a speaker that reproduces the sound waves. Figure 3.2 provides a depiction.

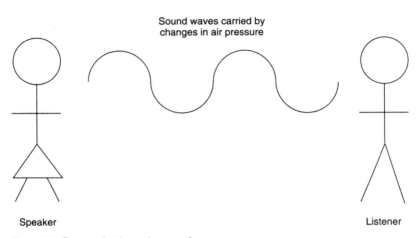

Figure 3.1 Communication using sound.

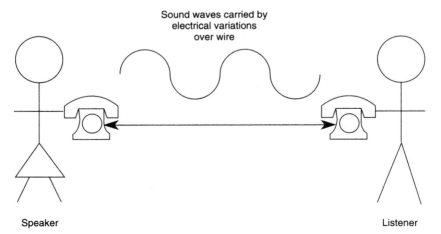

Figure 3.2 Communication using a telephone.

Discoveries made at the start of this century provided us with an even more sophisticated way to communicate information—using electromagnetic radiation. Electromagnetic radiation is another type of energy wave produced through the combination of a changing electrical charge and a magnetic field. This type of radiation includes radio and light waves, as well as other less familiar types of waves such as X-rays and gamma rays. What makes all these waves different from other types of radiation is that they do not require any material medium, such as copper wire, for transmission. That's why radio and light waves can travel across the void of outer space. And, of course, the lack of need for a material transmission medium is what makes them useful for wireless communication.

Key components

Regardless of the use made for wireless communication, the same two key components are required in order to convey information:

- A transmitter, which takes as its input some information like a human voice or a stream of data to be communicated. The transmitter creates waves that contain the information and outputs them using an appropriate device. For example, a radio transmitter uses an antenna to release its electromagnetic waves, while an infrared transmitter uses an infrared light-emitting diode or laser diode.

- A receiver, which captures electromagnetic waves and processes them to recover the information or data contained within. Its output is the original information that was input to the transmitter.

Because communication is normally a two-way street (ignoring, for the moment, the world of politics), each party to the process requires both a transmitter and receiver. The two units are often combined into one, known as a transceiver. Depending on the communications schemes used, only one party might be able to transmit at a time, or both parties might transmit and receive simultaneously (not unlike a debate in your average legislative assembly, except in this case, both ends actually listen as well as talk).

Figure 3.3 shows how transmitters and receivers process information. Note that the type of components used differs based on the type of electromagnetic radiation produced (for example, radio versus light). However, the principles remain the same. Obviously, transmitting and receiving electromagnetic radiation is the focus of wireless communication. Let's look a bit closer at this phenomenon.

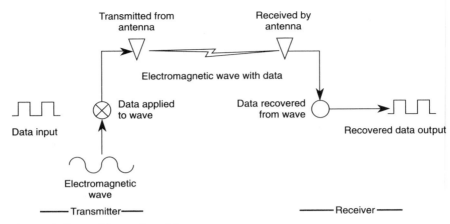

Figure 3.3 Transmitting and receiving.

More on Electromagnetic Radiation

Electromagnetic radiation includes many different types of waves: radio, light, and others. Waves are usually depicted as in Figure 3.4, which shows a graph of the amplitude or strength of the wave as a function of time. This results in a sinusoidal wave form, or "sine wave" for short. A cycle is from the top of one wave crest through to the next crest.

Aside from their amplitude, waves can be characterized in two main ways:

- Their frequency, which is the number of cycles they complete in one second. Frequency is measured in hertz, which equals one cycle per second. Prefixes such as mega (one million) and giga (one thousand million, or one billion) are used to indicate higher frequencies. For example, one megahertz indicates that the wave completes one million cycles every second. Megahertz is abbreviated as MHz, while gigahertz is abbreviated as GHz.

- Their wavelength, which is the distance between wave crests. Wavelength is inversely proportional to frequency, that is, higher frequencies have shorter wavelengths, while lower frequencies have higher wavelengths.

(For mathematics and physics buffs, the velocity of a wave is equal to the frequency times the wavelength. Thanks to Einstein and his peers, we know that the speed of all types of electromagnetic radiation through a vacuum is constant at 300,000 kilometers per second. As our algebra professors used to say, it will be left as an exercise to prove that frequency in megahertz equals 300 divided by wavelength in meters.)

Figure 3.5 shows the electromagnetic spectrum and where various types of waves fit in. As you can see, the full spectrum of electromagnetic radiation includes everything from longwave radio, through shortwave radio and television, into microwaves, through invisible and visible light, and onto X-rays and gamma rays.

All of these waves are electromagnetic radiation. However, as you can imagine, they have different characteristics and abilities for communication. Key considerations, such as how far the waves can go, what they can go through, and how much data they can carry, all have an impact on how useful they will be for different types of wireless communication. You'll look at some of these considerations in the next sections.

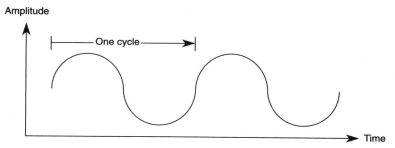

Figure 3.4 A typical electromagnetic wave.

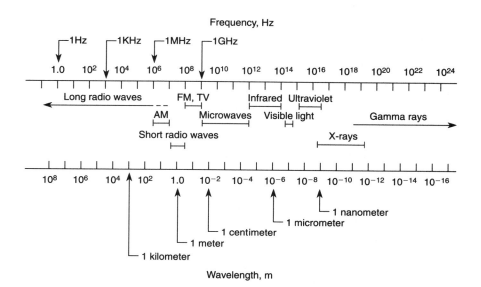

Figure 3.5 The electromagnetic spectrum.

Range of transmission

Whenever someone uses a radio, be it a "walkie-talkie," a mobile radio in a police car or taxi, or a cellular phone, somebody will always ask one key question: "How far can you get with that thing?" Of course, they want a nice, simple answer (the farther the better). But the communication range provided by a transmitter depends on many factors. As you'll see, the frequency of the wave is the key.

Like children in a car, different frequencies of electromagnetic radiation travel quite differently. None of them ask, "Are we there yet?," but some certainly go farther than others. For example, you might be familiar with what is commonly known as shortwave radio. The frequencies used range from about 3 MHz to 30 MHz and are also known as "high frequency" or HF. These frequencies are reflected by the earth's ionosphere, bouncing back and forth between it and the earth's surface. As a result, they can travel many thousands of miles and so are used by ham radio operators to talk to each other and by radio stations such as the Voice of America, the BBC, or Radio Canada International to broadcast programs to listeners around the world.

HF radio waves also are good at getting past obstructions like mountains, buildings, and people. While they lose intensity as a result of bouncing around or passing through the things that get in their way, the signal still can be usable over great distances.

Contrast shortwave with microwave transmissions. These much higher frequencies (more than 3 GHz, including "superhigh frequency" and "extremely high frequency") normally operate on a line-of-site basis. They do not "bend" around the earth's surface like HF waves. So, for example, telephone systems that use terrestrial microwave links require devices called repeaters every few tens of miles. The repeater contains both a receiver and a transmitter that operate simultaneously. The repeater receives signals from the last station (the original transmitter, or another repeater), amplifies them, and transmits them on to the next station. In some cases, short range is very desirable because it decreases the area of opportunity for someone to try to eavesdrop on transmissions.

Microwaves will go right through some objects. (Ask the last bag of popping corn you placed into your microwave oven if it didn't feel the sensation.) But microwaves are blocked by more solid objects like buildings or metal structures. The microwaves just bounce off solid objects such as these (giving rise to one of the less spectacular innovations of our time, the police radar gun).

Light waves behave in the same fashion as microwaves, but light waves are even more restricted in how far they will travel. And, of course, they have even less ability to penetrate objects, even a bag of popping corn. Again, this might be very desirable because it keeps your transmissions isolated, say within a single room. (Please note that the authors are not suggesting using bags of popping corn as security devices.)

But back to our friend who wants to know "how far." Well, there is even more to the question than a wave's frequency. The power of the transmitter, the sensitivity of the receiver, the kinds of antennae attached to each, the quality of the transmission line used, the height of the antennae, the method used to send information on the wave, the amount of interference present, atmospheric conditions, and even the point in the sunspot cycle can all have an effect on communication range.

Of course, for a wireless local area network, you are interested in a setup that keeps transmissions confined. That means techniques such as using higher frequencies, lower power, and antennae designed to limit distance.

Information-carrying capacity

The frequency used also impacts how much information or data can be carried on the wave. As you'll see in the next section, to convey information, electromagnetic radiation must be modified in some way, with the changes representing the details being transmitted. Generally speaking, the higher the frequency, the greater its capacity to carry information. This is because the information-related modifications made to the electromagnetic wave are normally tied to its frequency. As a result, HF radio waves have significantly less data-carrying capability than microwaves, which in turn have less capacity than light waves. (That is why fiber-optic cables, which use light to send data, can handle such large volumes, 500 to 1,200 Mbps.) The term *bandwidth* refers to this characteristic. Higher bandwidth means that more information can be sent in a given period of time.

The need for regulation

Some types of waves, particularly radio waves, are more likely to interfere with other services operating on the same frequencies. A transmitter operating in the HF spectrum could be heard around the world. Even shorter-range microwave transmissions operating on the same frequency must be kept separated by a reasonable distance. The potential for conflict and interference requires coordination between users, coordination that won't necessarily happen without the interference of regulatory boards and commissions.

Organizations such as the International Frequency Regulatory Board deal with those portions of the spectrum with the potential for worldwide interference. Regionally, agencies such as the Federal Communications Commission (FCC) in the United States and the members of Conférence Europénne des Postes et des Télécommunication (CEPT) are responsible for controlling specific portions of the spectrum within their countries.

Note that these agencies only concern themselves with the radio portion of the spectrum (under about 100 GHz), ignoring light-wave frequencies and above.

Control over the radio spectrum users means regulating those frequencies that people use, how much power is transmitted, and the physical configuration of the equipment and antennae. Ordinarily, this means that the regulatory agency issues a license to the user, with strict operating specifications that must be complied with at all times. This costs money. Applying for and acquiring the license, setting up an acceptable transmission system, maintaining compliance, and annual renewals all greatly add to the cost of operating a wireless communication facility.

However, some portions of the spectrum are set aside for use by transmitters with low power output (usually less than one watt). The nice part is that these frequencies can be used without licenses, as long as you use only authorized and unmodified equipment. For example, the specific design of a transmitter must receive "type approval" from the regulatory body before it can be sold, so manufacturers must prove that their equipment complies with the applicable regulations. Citizens band (CB) radio, popular in North America in the 1970s, was an example of such a service. Today's portable phones and room monitors also operate at low power, without the need for a license.

But the problem here is that interference is common, as you might know from hearing your neighbor's telephone conversation on your handset (which works both ways, you know). The trade-off is that, while you don't have to get a license, you might experience interference. So it seems like you are forced to choose between paying for dedicated spectrum space or sharing the space and getting unsatisfactory communication results.

Technology to the rescue. A transmission technique known as spread-spectrum technology provides a workable compromise between the need for dedicated frequency space and the havoc of shared services. You'll see more about spread-spectrum later in the chapter.

Equipment costs

One mathematical relationship might seem surprising. As the frequency goes up, so does the cost, at least within the radio portion of the spectrum. At higher microwave frequencies, electronic components must be higher quality and have better tolerances. Design factors like how the components are laid out become far more crucial. Even the wire that carries the signal to the transmitting antenna costs more. As a result, equipment operating at microwave frequencies tends to be more expensive than equipment using lower radio frequencies.

Fortunately, as has happened for most electronic items, costs continue to decline. So microwave systems, which have higher data capacities, are becoming more and more common.

One area where the cost/frequency equation falls down is with light-based systems. Despite the fact that they are much higher frequencies than microwave, systems such as those using infrared transmitters are surprisingly inexpensive. This owes to the different and cheaper technologies used to produce light transmissions. A common example is the inexpensive light-emitting diode (LED), which often serves as the transmission source.

Adding the information—modulation

So far, you have learned about electromagnetic radiation, some of its characteristics, and how it is transmitted and received. But how does the information you want to convey, whether it is the human voice or a data file on a personal computer, get sent on the electromagnetic wave?

The answer is that the transmitter must apply the information to a basic wave (known as the carrier) in such a way that it can be recovered by the receiver. In its simplest form, this can mean starting and stopping the carrier in a specific sequence. For example, Morse code can be sent by simply turning the carrier wave on and off to form short and long pulses (dots and dashes).

More sophisticated methods involve altering a basic characteristic of the carrier wave. This process, known as modulation, can be performed in different ways, two of which you have probably heard of:

- Amplitude modulation (AM). An audio signal like the electrical output of a microphone is used to vary the intensity (known as the amplitude) of the carrier. The receiver detects the changes in amplitude and recreates the audio output.

- Frequency modulation (FM). The same electrical audio signal is used to vary the frequency of the carrier. Circuitry in the receiver discriminates between the varying frequencies received to recover the audio information.

Figure 3.6 shows the sine wave patterns resulting from AM and FM.

Other modulation schemes include such technical-sounding names as phase modulation, pulse modulation, on-off keying, frequency shift keying, phase shift keying, and (everyone's favorite) quadrature amplitude modulation.

Of course, how the modulation scheme works probably is not important to you as a user, as long as it does the job of reliably transferring information over the carrier wave. The wireless LAN manufacturer will have to worry about making that happen.

Sharing the spectrum

Earlier in the chapter, you learned about the problems associated with sharing a band of frequencies with other unlicensed users: conflicts and inter-

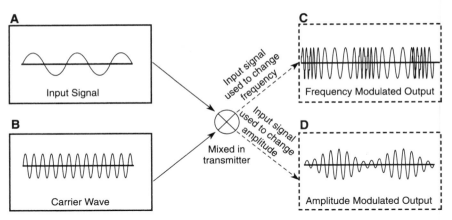

Figure 3.6 The effect of modulation.

ference. If another user has the same type of communication equipment as you, then there is definitely a chance that your operations will interfere with each other.

Fortunately, a technique pioneered by the military to help secure its transmissions offers a way around this problem. Known as spread-spectrum technology (SST), it involves ways of spreading transmissions across a range of frequencies, rather than transmitting on one frequency all the time. Figure 3.7 illustrates how a spread-spectrum signal compares with a conventional signal.

One approach to spectrum spreading, known as frequency-hopping spread spectrum (FHSS), involves dividing a range of the radio spectrum into individual channels, each on one specific frequency. A transmitter can use any subset of these channels. The transmitter will hop from channel to channel during its transmission, following a specific pattern. Using a mechanism such as the Code Division Multiple Access (CDMA) protocol, the receiver must follow the transmitter as it hops between frequencies.

At the same time as one transmitter/receiver pair hops around the band, other transmitters and receivers are doing the same thing across the same group of channels. Everyone shares the same band of frequencies, but the SST greatly diminishes the amount of interference.

Another method of spectrum spreading is known as direct sequence spread spectrum (DSSS). Here, transmissions occur not on one channel at a time, but over a continuous band of frequencies. A unique spreading code is applied to the input signal before it is applied to the carrier wave. The receiver uses the same code to recover the information.

SST is an important basis for many wireless LANs, especially those found within a shared portion of the ultra high frequency (UHF) range around 900 MHz.

Common Frequencies Used for Wireless LANs

Wireless LANs can be found in two main areas of the electromagnetic spectrum:

- Radio LANs, currently operating in the high UHF and low microwave range.
- Infrared LANs, using transmissions just below visible light.

Because infrared LANs use light with wavelengths of less than 1 micrometer, they are not subject to regulation by groups such as the FCC and CEPT. Radio LANs are a different matter. Obtaining space for radio LANs has not been easy. Overcrowding already exists in the radio spectrum, and existing users are battling to maintain their space. They do not look kindly on new services that take up valuable spectrum, particularly in the low microwave range.

So far, some progress has been made for radio LANs. Frequencies in the high UHF, low microwave range provide a good combination of characteristics such as limited distance (so that your LAN doesn't interfere with others) and the ability to go through walls and floors (for use between a number of offices). So in 1985, the FCC in the United States allowed unlicensed wireless LANs to operate in three areas of the radio spectrum known as the industrial, scientific and medical (ISM) bands:

- 902 to 928 MHz.
- 2.4000 to 2.4835 GHz.
- 5.7250 to 5.825 GHz.

It's not only wireless LANs that use this space. For example, the 902-MHz portion also is available to baby monitors and cordless telephones. For

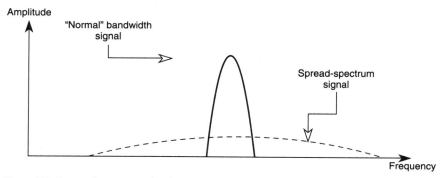

Figure 3.7 A spread-spectrum signal.

LANs to use this part of the spectrum, they must use the space without disturbing existing users, and with no protection against disturbances from existing users. As a result, wireless LANs need to use techniques such as SST, not just to prevent interference from other LANs, but to avoid increased data errors when Junior decides to wake up for a feeding.

As mentioned in chapter 1, the fight for frequencies will continue for some time to come. For example, the FCC is considering allocations in the 17-GHz and 61-GHz bands. The FCC also allocated 1.890- to 1.930-GHz for use by unlicensed personal communications services (PCS). PCSs include both voice and data wireless services, as manufacturers from each camp want more space for their use. But no one wants to incur the majority of costs for relocating existing users to other portions of the spectrum, a condition for getting the space.

The European Telecommunication Standards Institute (ETSI) along with the CEPT has allocated 1.880 to 1.900 GHz as part of the Digital European Cordless Telecommunications (DECT) standard. While the entire 2.4-GHz band used for ISM in the United States is unavailable, the CEPT also suggests that 2.445 to 2.475 GHz be made available for low-power SST-based wireless LANs. Bands in the 5-, 17-, 18-, and 24-GHz ranges are also being considered.

Manufacturers of wireless LANs are the ones who have to worry about frequencies they can operate on. Unless you use a licensed service, you pretty much just install the LAN and turn it on. However, it helps to be aware of what frequency band a wireless LAN operates on because it affects how well it might perform in your circumstances, or what sources of interference it might face.

Wireless Architectures

In the last chapter, you learned about local area networks and how they operate. After getting this far in the chapter, you have a better understanding of how wireless communications work. Now we can look at how communications technology is applied to construct wireless LANs.

Basically, wireless links fill in for wired links using electromagnetic radiation transmitted at radio or light frequencies between transceivers. A wireless link can:

- Replace a point-to-point connection between two nodes or between segments on a LAN.

- Provide a connection to a wired LAN on behalf of one or more wireless nodes.

- Act as a stand-alone LAN for a group of wireless nodes.

Of course, there are variations on these themes, but they illustrate the basic building blocks used. Let's look at each of these examples more closely.

Replacement of a point-to-point link

A point-to-point link is a connection between two devices used to transfer data. Examples include the wire from a terminal to a mainframe, from a PC to a printer or modem, or from a desktop PC to a notebook computer. Normally, the two devices send data back and forth, following rules that specify who can send at a given point in time.

Figure 3.8 shows a point-to-point link using a wire between two computers. It also shows how the link could be replaced by a wireless link. Instead of a wire running all the way between the machines, each computer is connected to an infrared transceiver. Each transceiver is pointed towards the other so that the light passes between them unobstructed. The transmitter at each end applies the data sent by its computer to a light signal, which is received at the other end. The data is then recovered from the signal and passed to the receiving computer.

This is the simplest form of a wireless network because it only involves two devices. Nevertheless, it shows that, in many cases, a wireless link is simply a direct replacement for a wire connection.

A wireless link can be used to bridge two LAN segments. Recall from chapter 2 that a bridge is an internetworking device. It looks at the network traffic on each of the segments it is connected to. If it sees a packet on one segment addressed to a node on the other segment, it forwards the data. Figure 3.9 shows a wireless bridge. Similar to a point-to-point link, it simply connects bridging devices attached to each LAN, allowing packets to be forwarded across the wireless link. This configuration is quite useful for linking LANs in two buildings that cannot be easily wired together directly (for example, two office buildings separated by a street or highway.)

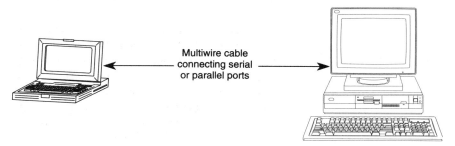

Multiwire cable connecting serial or parallel ports

Figure 3.8 Point-to-point link using a wire.

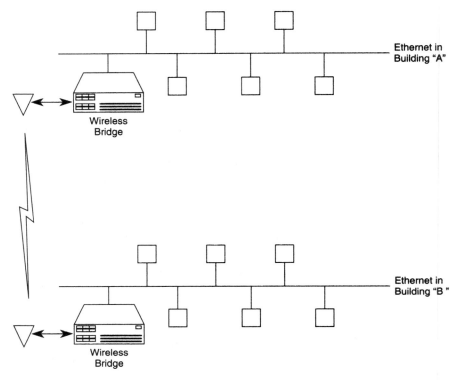

Figure 3.9 A wireless bridge.

Connection of wireless nodes to a wired LAN

As you'll see later in this book, it is very common to integrate a wireless LAN with a wired LAN. This often occurs where there is an existing wired LAN coupled with a need for some wireless connections (for example, with notebook computers that are used at various places around an office building).

A typical approach is to attach a device to the wired LAN, as shown in Figure 3.10. The device acts as the point of contact between the wired LAN and the wireless nodes. Several different techniques might be used to transfer data between the wireless nodes and the wired network. For example, the wired and wireless portions might be viewed as separate LAN segments. Here, the device can act like a bridge or router, passing the appropriate traffic between each segment. Another approach is that the device repeats everything it sees on the wired LAN for the benefit of the wireless nodes.

Stand-alone wireless LAN

Given the right circumstances, wireless nodes are perfectly capable of forming their own stand-alone wireless LAN, with no connections to any wired LANs. All that is required is for each node to be able to send and receive messages to and from other nodes. This can be done using architectures that are very similar to the wired architectures discussed in chapter 2. For example:

- A star can be formed where each wireless node communicates directly with a central hub, which forwards the message on to the intended recipient.

- A ring can be used, where each wireless node sends data to its neighbor, which receives messages intended for it or passes it on to the next node.

- A bus exists when each wireless node can hear everything said by all the other nodes.

Figure 3.11 shows how these can be configured.

The choice of architecture depends, to some extent, on the type of wireless technology used. For example, an SST radio LAN is normally a good choice for a bus because the electromagnetic waves will spread out over an area and allow each node to hear every other node. An infrared signal is far more directional and is therefore suited to having each transceiver point to a central hub, as in a star. However, infrared transceivers can also be

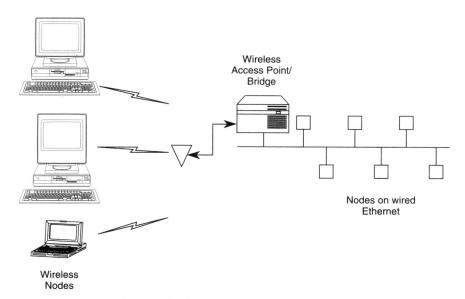

Figure 3.10 Interface of wireless LAN to a wired LAN.

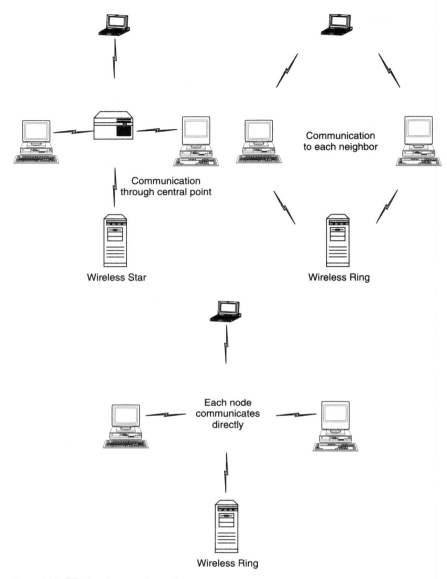

Figure 3.11 Wireless bus configurations.

pointed at a common reflector (for example, at the mid-point of a ceiling in a room). If all transceivers point and reflect at the same location, a bus could also be achieved, as shown in Figure 3.12. A combined approach could involve a central hub for all wireless nodes, which also acts as a bridge connected to a wireless LAN, as shown in Figure 3.13.

So far, this book equates connection and communication, with a reference to the need for a common language. Perhaps an anecdote is appropriate here to demonstrate a fundamental rule of networking: connectivity doesn't necessarily mean communication. There's a story making its way around Toronto involving a trucking company and its new wireless data communications system. The satellite-based system allows the dispatcher constant and immediate contact with drivers on the road. The company believes the new system will increase efficiency and productivity, primarily by eliminating dead-runs.

Well, it seems that one day a message came in from a driver who had just unloaded his truck in Chicago: "Anything to pick up before I come back?" Since there wasn't a load for pickup, the dispatcher sadly responded, "No, come home."

The dispatcher was worried the next day when the truck did not return to the company's terminal. Anxious, the dispatcher sent off the message, "Where are you?" Imagine the dispatcher's surprise when the message "In Chicago," came back. At which point, the dispatcher typed "Why?" "You said 'no come home,' so I didn't," returned the driver. This simple story illustrates that connectivity doesn't mean communication. When you have

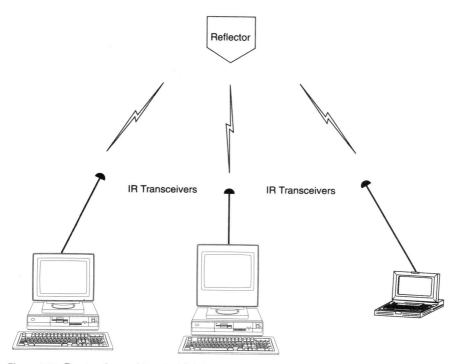

Figure 3.12 Bus topology achieved with infrared technology.

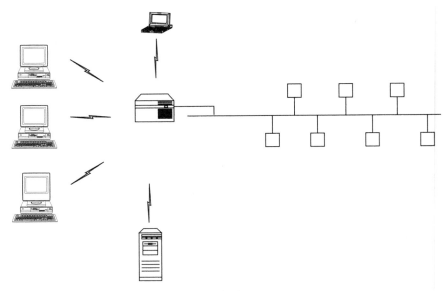

Figure 3.13 Wireless access point serving as a bridge.

problems as simple as this, then wireless won't help! But for the rest of you, wireless LANs might be the answer.

Hopefully you now have a better appreciation for how wireless LANs function, the common frequencies and techniques used, and some of the terminology you might run across. In the next chapter, you'll read about the standards being developed to help ensure interoperability between wireless LANs. In chapters 6 and 7, you'll see some of the practical applications for wireless LANs in your business.

4

Standards for Wireless LANs

Standards can make your life a little less hectic and unpredictable. Whether it's the side of the road you drive on, the quality of meat or vegetables you buy, or the training of medical staff in your state or province, standards can give you one less thing to worry about as you make your way through daily life.

The world of computers and networks also benefits from standards. For example, a standard has evolved for personal computers based on the Intel 80x86 central processing unit (even though Macintosh users might disagree). PCs using these chips are found all over the world. They run the same programs and read and write the same kind of data. So you can get comfort from knowing that one machine will read the disk from another (although sometimes it won't), and that one modem will talk with another (although sometimes they don't), or that one word processor will understand a file created by another (although they only seem to every so often, when the stars are properly aligned).

Ahem. Well, despite problems that crop up, standards for computers, and in particular for computer networks, have come a long way in a relatively short period of time. And the development of open standards for local area networks has helped them gain acceptance for use in business-crucial information-processing systems.

Standards for wireless LANs are still in their infancy, not unlike those for wired LANs ten years ago. However, progress is being made, and wireless LANs might soon also be accepted as a routine component for business systems. You'll learn more about the current state of standards for wireless LANs in the following sections.

The Need for Standards

During the formative years of the computer industry, there were not many standards accepted by all vendors. Instead, manufacturers designed their own machines and software from the ground up. Virtually every aspect of the computer, from the CPU, to the disk drives, to the operating system, were totally proprietary to that manufacturer.

Proprietary solutions have some benefits. They can be very efficient (just like a dictatorship). Because manufacturers don't need to consult with other members of the industry, they can do whatever they want with their hardware or software specifications. And while nonproprietary standards are supposed to allow hardware and software from different vendors to work together without any problems, it doesn't necessarily work that way. Products from one source based on proprietary specifications will usually have a better chance of working together (although this also isn't necessarily so).

However, proprietary solutions suffer from two big disadvantages:

• Their components tend to be very parochial, not sharing information or facilities with equipment from other manufacturers, or perhaps not even speaking the same language, such as when using differing data formats or network protocols. In addition, some people view the development of proprietary solutions to be a less creative process, which can fall prey to the inbreeding of ideas.

• Because the solutions are proprietary, the specifications are generally kept exclusive. Other manufacturers cannot produce competing products, or they have to pay large licensing fees to produce components that comply with the specifications. This keeps the prices for equipment up, which is good for the proprietary vendor but bad for its customers.

So while proprietary solutions offer more direct, hassle-free systems, they also are less flexible and often cost much more. But they worked in the past, and data processing managers continued to not be fired for buying computers from the traditional suppliers (for example, Big Blue). Proprietary vendors became blasé, counting on their customers being locked-in and resting on past solutions and high profit margins.

Fortunately, with the advent of the personal computer dawned an age of more open standards. For example, while it almost certainly wasn't the intention of IBM, after it introduced the IBM PC in 1981, more and more clone or knock-off computers became available that would run the same programs. Of course, the clones and knock-offs were less expensive, with new manufacturers undercutting those who had previously undercut their rivals. And the prices just keep on falling.

The result is today's personal computers, running tens or hundreds of times faster than the original PC, with thousands of quality software titles. Today, you use diskettes to swap megabytes of data between machines more easily than placing a music cassette in different tape decks.

But its not just that we can run the same programs or exchange data between machines more easily. Today, standards like the Industry Standard Architecture (ISA) bus, and the Small Computer System Interface (SCSI) for personal computers allow hardware components from different manufacturers to be tied together within the same computer. And with an operating environment like Windows, data can be cut and pasted, or data objects shared, between software packages from different vendors.

Computer networks also have benefited from this same trend. For example, the Transmission Control Protocol/Internet Protocol (TCP/IP) is an open networking standard. Anyone can produce programs that comply with the standards, running on any machine of their choosing. TCP/IP is the basis for the worldwide Internet, which currently boasts anywhere between 2 and 3 million interconnected hosts, ranging in size from PCs and Macintoshes to IBM mainframes and beyond.

More directly related to standards for wireless LANs is the 802 Standards Project undertaken by the Institute of Electrical and Electronic Engineers (IEEE). The project has already produced specifications for wired local area networks, known as 802.3 (colloquially known as Ethernet, but not exactly the same), 802.4 (using a token-passing bus architecture) and 802.5 (using a token-passing ring). Components complying with the appropriate standards can function properly with other components, even if they are from a different manufacturer. As a result, you can build wired LANs using network interface cards from different vendors. You can choose cards based on price, on features, on vendor reputation, but you don't have to buy them all from the same place. (You'll see more about standards from the IEEE, as well as other organizations a bit later in this chapter).

Standardization played an important role in the acceptance of wired LANs, and it will likely do the same for wireless LANs. Unfortunately, as has happened in many areas of computing and electronics, standards are not as standard as they might seem. (Remember beta-format video cassette recorders?) To paraphrase a well-known quotation on the subject: Standards are great. That's why there are so many to choose from. Fortunately, some progress towards one standard is being made, as you'll see in the next sections.

Where the Standards Apply

To understand the kind of standards developed for wireless LANs, it helps to bring in some frame of reference for the network communication process.

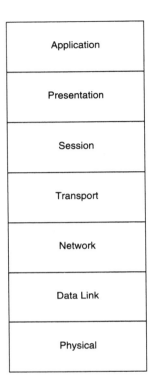

Application
Presentation
Session
Transport
Network
Data Link
Physical

Figure 4.1 The OSI Model.

Recall from chapter 2 the discussion of the OSI Model for network communications. The ISO developed the OSI Model to illustrate the separable functions needed for data communication between computers (see Figure 4.1). So let's have a closer look at the current state of standards, starting with how they stack up against the OSI Model.

The affected OSI layers

The bottom two layers are the ones of interest to us for wireless LANs because they define the physical characteristics and access rules for the network.

At the very bottom is the Physical Layer. This layer defines the electrical characteristics of the actual connection between network nodes. For wired LANs, it covers topics like voltage levels and type of cabling. But for wireless LANs, it addresses areas such as frequencies used and modulation techniques, including, for example, use of spread-spectrum technologies.

The next layer up, known as the Data Link Layer, deals with how the network is shared between nodes. The Data Link Layer defines rules such as who can talk on the network, how much they can say, and how they know if their message got through. One popular set of standards defines this layer as two separate layers:

- The Medium Access Control (MAC) layer, which sets rules covering when each node on the network can send messages.
- The Logical Link Control (LLC) layer, which provides a connection-oriented service (a logical link) between nodes.

See Figure 4.2 for an illustration of these layers and sublayers.

Techniques such as those mentioned in chapter 2 must be used so that more than one node on the network doesn't try to send traffic at the same time. Contention schemes such as Carrier Sense Multiple Access with Collision Detection (CSMA/CD) or Collision Avoidance (CSMA/CA), or deterministic methods such as token passing, are as equally valid and necessary for wireless LANs as they are for their wired counterparts. As we review the current standards, you'll see that the focus of standard setting for wireless LANs has been on the Physical and Data Link layers.

Effect on upper layers of OSI Model

In a perfect world, or at least an OSI-compliant world, each layer of the model works away on its own, interacting only with the layer above and the layer below, by using a standard dialogue. If everything works as planned, each layer really doesn't need to know the details about what's happening in the layers below it or above it. Ideally, then, we could simply substitute a wireless LAN for a wired LAN by inserting wireless components at the two lowest layers.

Unfortunately, some complications do occur and affect the operations of higher layers. For example, moveable devices like a wireless LAN node have a habit of temporarily dropping out of site, perhaps reappearing somewhere else on the network. This can confuse higher-level functions like routers and network operating systems, which expect devices to stay put once they show up on the network. Another example is the longer delays that can be found when using a wireless LAN, due to the generally slower data transfer speeds (particularly for radio LANs). Network pro-

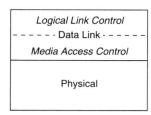

Figure 4.2 OSI layers covered by the IEEE 802.11 standard.

grams that expect responses within a certain time period might have to be adjusted accordingly. So functions operating at the upper layers of the OSI Model might actually have to take account of the fact that wireless technology is being used.

The Current Standards

To this point in the development of wireless LANs, two approaches have contributed to the development of standards:

- Formal standards-setting processes by groups such as the IEEE and the European Telecommunications Standards Institute (ETSI).
- Product development by individual manufacturers trying to sell various types of wireless LAN components.

The standards-setting process is a lengthy one, involving consultations, meetings, proposals, and counter proposals. As with any type of committee work, differences of opinions can range from minor disagreements on procedures to near religious fervor regarding basic philosophical differences. It's no wonder, then, that wireless standards have been slow to emerge.

Meanwhile, vendors (who only have to face internal committees) have brought out many wireless LAN products even without the benefit of standards. Appendix A provides you with details on many of the current products available.

Because most of the manufacturers involved with wireless LANs are also represented on committees such as the IEEE, the two approaches aren't really separate. Often, one manufacturer can propose its own products as the basis for an open standard. Or a group of manufacturers might work together to develop a specification.

This might seem like it surrenders a competitive advantage. After all, why make the fruits of your internal development efforts available to other companies in the same market? However, with the current demand for nonproprietary, interoperable systems, it actually makes sense to try to get your specifications adopted as the standard. After all, you'll be the first one out with a compliant product. And the product will be more attractive because it interoperates with others.

The IEEE has had great success with their wired LAN standards. For example, the ISO and its International Electrotechnical Commission (IEC) adopted many of the IEEE's 802.x standards internationally and will consider the Institute's work on wireless LANs, once standards are complete. So it is likely that the standard with the greatest impact will be from the IEEE's Working Group P802.11, charged with the responsibility for drafting wireless LAN standards. We'll consider these standards in the next section.

IEEE 802.11 standard for wireless LANs

Originally formed in 1990, the P802.11 Working Group took on the task of developing standards for all kinds of wireless communications. Their efforts gradually became more focused on individual areas such as wireless LANs, and on specific components of a wireless LAN standard. As a result, in November, 1993, the first portion of a wireless standard was approved, known as the "foundation protocol." Work on other portions continues to progress.

MAC level. The foundation protocol is at the Medium Access Control level of the Data Link Layer. Remember that this specifies the nodes that can access the transmission medium at any point in time.

The MAC standard will allow networks with more than 1,000 nodes and will handle data transmission speeds up to 20 Mbps. Of course, the actual speed will depend on the Physical Layer protocol in use. You'll see more about that a bit later.

The MAC protocol arrived at for wireless LANs uses Carrier Sense Multiple Access with Collision Avoidance (CSMA/CA). When using CSMA/CA, each wireless node with data to send waits for silence on the network before it tires to transmit. When silence is detected, the node transmits its data, then waits to receive a formal acknowledgment of its transmission from the recipient node. If it does not receive an acknowledgment in a set period of time, it assumes that a collision has occurred. A collision means that another node has transmitted at the same time. If this happens, it is likely that the two messages have become garbled, and that the data is lost to the recipients. Each node then waits for silence, then delays its transmission a randomly generated further period of time. The random delay is designed to prevent yet another collision between the two nodes trying to send data. If it still hears nothing, the node tries transmitting again.

This process contrasts with CSMA/CD. Here, when a node has data to send, it also waits until no other stations are heard on the network. But it also listens to hear whether any other nodes also are transmitting. If it hears another node, it assumes that a collision has occurred, and follows a process similar to CSMA/CA for trying again. For example, wired LANs that follow the IEEE 802.3 standard use collision detection.

While collision avoidance protocols are slower than collision detection, due to the need for formal acknowledgments, they work well for wireless LANs. A node on a wireless LAN might not always hear every other node, so it might not detect every collision that can garble its message at the receiving end. So it's better to get an explicit acknowledgment from the recipient, rather than assuming that the message made it through.

One benefit of the 802.11 MAC-layer protocol is that it is tied to the IEEE's 802.2 Logical Link Control layer (as are the MAC-layers of the other

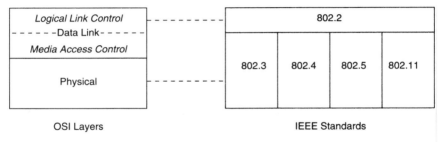

Figure 4.3 Relationship of lower layers to the 802.2 LLC standard.

802.x wired LAN standards). This makes 802.11-compliant LANs easier to integrate with other LANs also conforming to the 802.2 LLC standard. Figure 4.3 illustrates this.

The acceptance of the MAC-level standards paves the way for vendors to market compliant products, although this might take until late 1995. However, standards compliance makes the products more attractive, since it brings the benefits of standardization like products from different vendors that will work together. However, the MAC-level is only one portion of a wireless LAN standard. There are still the lower-level Physical Layer specifications that must be addressed.

Physical Layer. The foundation MAC layer must be supported by standards for the Physical Layer. As you might expect, each kind of LAN covered in chapter 3 requires unique specifications of how data is transmitted. So at the physical level, work continues on three different sets of specifications for:

- Infrared light LANs.

- Frequency-hopping spread-spectrum (FHSS) LANs.

- Direct sequence spread-spectrum (DSSS) LANs.

Figure 4.4 shows how the IEEE physical protocols fit together with the MAC-level.

Some progress has been made in these areas. For example, in the area of FHSS LANs, specifications have been set for 1-Mbps transmission speed. The same speed likely will be specified for DSSS LANs. In addition, for FHSS LANs, a faster specification allows 2 Mbps, with provision for a fall back to 1 Mbps if necessary for compatibility and reliability.

Although still not complete, the IEEE standards very likely will play a dominant role in the final wireless LAN standards used worldwide, if past history for wired LANs is any indicator. But the worldwide acceptance of the IEEE's wireless standards faces a tougher challenge—coordination for the use of radio frequencies, at least for spread-spectrum LANs. Coordination means acceptance by groups such as the European Radio Council (ERC), as well as organizations in individual countries, like the U.S. Federal Communications Commission (FCC).

The European Telecommunication Standards Institute

On the European continent, the ERC allocates portions of the radio spectrum for use by the 32 member countries. Standards for use of allocated spectrum are then created by the European Telecommunication Standards Institute (ETSI). The ETSI has turned its attention to standards for radio frequency wireless LANs. And because wireless LANs could play an important role for data communications and networking in Europe (particularly in Eastern Europe, where there is not an extensive wired infrastructure) the ETSI standards will figure prominently in their future development.

As mentioned in chapter 3, a 20-MHz portion of the microwave spectrum was allocated for use by the Digital European Cordless Telecommunications (DECT) standard in March, 1992. The DECT standard deals not only with wireless LANs, but it also covers other wireless applications like voice communication over wireless telephone exchanges. But it can be used for data as well, and some vendors offer wireless LAN products that operate in the 1.88- to 1.90-GHz range of the DECT band. The DECT standard defines protocols for both the Physical and Data Link Layers of the OSI Model, as shown in Figure 4.5.

However, the ETSI is also close to finalizing its standard known as RES.2. The standard is based on the 2.4 GHz range approved for use by low-power, spread-spectrum LANs in many European countries. This might prove to be a very useful specification that gains worldwide acceptance, should it receive a positive reception in the European Economic Community and other European countries.

Future Standards Development

While the future is difficult to predict, based on past trends, the importance placed on interoperability and nonproprietary solutions will continue to

Media Access Control		
Infrared Physical Layer	FHSS Physical Layer	DSSS Physical Layer

Figure 4.4 IEEE 802.11 Physical Layers with the MAC foundation protocol.

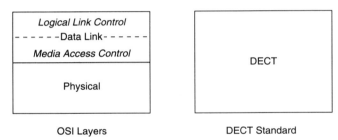

Figure 4.5 OSI layers covered by the DECT standard.

drive the standard-setting process. It might be slower for wireless LANs than their wired counterparts, owing to the additional complications of frequency allocation. But it will happen.

The IEEE 802.11 committee currently schedules the standard for full completion by March, 1995. Meanwhile, approval for the ETSI RES.2 standard is imminent. The ETSI also is working on RES.10, a standard for High Performance European Radio LANs (HIPERLAN), which will make use of space in the shorter microwave bands (5.7 GHz and 18 GHz) and operate at speeds up to 20 Mbps.

There also is talk of a joint ETSI/IEEE specification for high-performance LANs operating at high microwave frequencies like 60 GHz and providing data transmission speeds of hundreds of megabits per second. So the standards, they are coming, but as our colleagues often say about us, "They're not all there" (at least not just yet).

The Impact of Standards on Current Wireless LANs

Because standards for wireless LANs still are under development, the question must be asked, "Should I hold off on installing wireless LANs until the standards are finalized?" Because standards-compliant components are preferable, surely it's better to wait.

But wireless LANs are like most kinds of computer hardware. There always will be something newer, better, faster, or cheaper on the horizon. While standards-compliant wireless LAN components will offer many advantages (including interoperability, and quite likely lower price), they aren't here yet. However, there might be an existing product which fills an existing business need very nicely. It might not integrate as easily with components from other manufacturers offered in the future, but that isn't necessarily a key criterion for the purchase decision today.

Basically, it all boils down to the same criteria as most other business decisions: cost versus benefits. Compare the cost of a current wireless solution (including hardware, software, ongoing support, and maintenance) with the potential benefits it can bring (including savings in money, savings in time, improved productivity or efficiency). If the cost is less than the benefits, it makes sense to invest in wireless now, even without a full set of standards.

Later, as the standards are finalized and compliant hardware becomes more common, you can consider the costs versus benefits of replacing your wireless LAN with up-to-date components. But in the meantime, your business has enjoyed the benefits of wireless technology. In the next chapter, you'll see some of the ways you can use wireless LANs to collect and deliver important information for your business.

The Needs

"A world community can exist only with world communication, which means something more than extensive shortwave facilities scattered about the globe. It means common understanding, a common tradition, common ideas, and common ideals."

Robert M. Hutchins

"There are two kinds of fools. One says, 'This is old, therefore it is good.' The other says, 'This is new, therefore it is better."

Dean William Ralph Inge

"There is one thing stronger than all the armies in the world, and that is an idea whose time has come."

Victor Hugo

5

Uses for Wireless LANs

You now have a better understanding of how wireless LANs function and the standards evolving around the world. Well, that's all very interesting, but it doesn't answer some key questions. For example:

- How do you use wireless LANs in real-world business information systems?

- Do they replace wired LANs?

- Can they work with wired LANs?

- Do they offer any new tricks, or just the same old functions?

The answers to these and other questions are in the following sections, which describe some of the ways you might incorporate wireless LANs into your computerized applications.

Comparing Wires with Wireless

Wireless LANs might seem like a direct replacement to wired LANs. After all, why go to all the trouble and expense of pulling cables when you can simply connect your wireless LAN card or adapter to your computer and begin networking?

It's true that, in some cases, a wireless LAN will serve as a direct replacement for wired approaches. If you're starting from scratch in building a brand new LAN, you should definitely consider wireless as an alternative. However, in the large majority of cases, a wireless LAN will complement a wired LAN, not replace it. There are two main reasons why wireless LANs won't just move in and take over from wired LANs: performance and cost.

Performance

In a direct comparison of data transfer speeds, a typical wired LANs will win out over wireless almost every time. While today's spread-spectrum LANs offer throughput of 1 or 2 Mbps, an IEEE 802.3 (Ethernet) LAN runs at 10 Mbps, while an 802.5 (Token Ring) runs at 16 Mbps. The comparison is somewhat apples-to-oranges, but it illustrates the magnitude of the difference.

Of course, other kinds of wireless LANs (like infrared) can provide better throughput that approaches speeds of wired LANs. And data speeds for radio frequency wireless LANs will inevitably increase as technologies improve and higher microwave frequencies are explored. But wired LANs also will get faster. Speeds of 100 Mbps over conventional wires soon will be common. Only specialized wireless applications (such as point-to-point laser links) can begin to approach that kind of speed.

In addition, wireless LANs are normally slower because they are less efficient. Because wireless communications are more prone to certain types of interference than wired links, they are also more prone to errors. As a result, wireless data transmissions must include information like additional error-correcting codes. Recall the use of CSMA/CA, which requires that acknowledgments are sent for every single message. These additional overheads to the communication process take up bandwidth and reduce effective throughput.

For performance, wires will always have the advantage. Even when the wire is actually a fiber-optic cable, it will still provide better performance than light transmitted directly. So wireless LANs will not replace wired LANs based on speed alone, at least not for the foreseeable future.

Cost

Common sense tells us that wireless LANs always will cost more than wired LANs. The technology is more complex and involves more components. For radio LANs, frequency spectrum is a limited resource and therefore will cost more to use. This cost is direct, like licensing fees, or indirect, through development and use of technologies like spread-spectrum to automatically share space (and therefore limit performance, when compared to dedicated frequencies). The price of today's wireless LAN products is certainly higher than an average Ethernet card. You need look no further than comparing the cost of using cellular telephones versus regular land-line telephones to have this borne out.

Perhaps surprisingly, there is some debate as to whether going with a wireless LAN can actually cost less. How can this be? The additional cost for moving cables with wired LANs becomes vividly apparent the first time you need to move people around your office. Removing old cables, drilling holes, disassembling or moving furniture, pulling new cables, and attaching connectors. All of this can be enormously expensive in materials and labor. With a wireless LAN, you just pick up and move, without incurring these costs.

Some organizations address the problem of relocating wires by pulling cables to all likely work locations right from the start. The theory is that a move simply involves disconnecting from one cable and attaching to another at the new location. (It might also be necessary to swap cables in a wiring closet, or at a hub site). This approach saves the cost of moving connections, but there is still a price to pay. Cables must be routed to locations that do not always need access to the LAN. In addition, the price, which still includes labor and materials, is paid in advance, rather than at the time of an actual move. (The only time this doesn't cost more is when everyone in the office requires LAN access. In that case, an office move is simply a case of musical chairs. When the music stops, everyone must grab a LAN connection, and the person left out must convert to "SneakerNET").

Different estimates float around about how much an average wired LAN node costs versus a node for an average wireless LAN. But the numbers presented are based on differing assumptions, such as the frequency of relocation, difficulty in rewiring, and the cost of labor. Results can be created to support both sides of the debate, meaning the inputs are subject to finessing. It's like the accountant who was asked by the owner of a company what her net profit was for the year: "What would you like it to be?"

In reality, there is no such thing as an average LAN node. The cost of installing and maintaining a wired LAN in your office or factory will vary widely, and it is your specific costs that need to be considered when comparing with wireless alternatives. However, if you often relocate LAN users and their connections, a wireless solution is definitely worth considering.

Integrating the Technologies

The most likely scenario is that wireless LANs will be integrated with wired LANs. This approach makes sense, for two main reasons:

- Many companies already have wired LANs, and in some cases, an extensive wiring infrastructure to support current and future needs. The old saying is, "If it ain't broke, don't fix it," and this applies to LANs as well. So, there really is no need to throw away wired nodes and replace them with wireless just because it's the latest and the greatest, or because you need a few wireless nodes. Expect to keep those wired nodes around.

- A wireless LAN connecting a few computers together to form an isolated group might make sense all on its own. For example, a small work group that needs to share files or a printer might find a simple stand-alone wireless LAN to be a convenient and cost-effective solution, especially if the group moves around a great deal (like a team of auditors). But in a world of distributed processing, client/server environments, and corporate electronic mail, connectivity is key. You already spend time integrating different types of wired networks together for greater access to resources: IEEE 802.3 with 802.5, IPX/SPX with AppleTalk, TCP/IP with SNA. The list of networks (and acronyms) seems endless. So, it probably doesn't come as a surprise that wireless network users will still want access to the wired infrastructure.

In addition, some companies are looking at wireless technology to connect hard-to-wire nodes such as an information kiosk in the atrium of a hotel.

As a result, most wireless LANs will have some type of connection to an existing wired LAN. This gives wireless nodes access to corporate resources already available to wired nodes, like file servers, printers, and electronic mail gateways. However, it's not only the wireless nodes that gain from this arrangement. Personal computers connected through wireless nodes can supply data to wired nodes in real-time, data that was previously unavailable or greatly delayed. Both wired and wireless users can benefit from being able to share data more easily. Figure 5.1 shows a connection between a wireless work group and a wired LAN.

The case for integrating wired and wireless LANs is similar to the telephone system since the advent of cellular telephones. Cellular phones certainly haven't replaced wired phones. Their initial cost has been much higher, and the price of air time adds up to large amounts very quickly. But cellular phones have filled a great need for temporary communication, or for mobile use, and so their use has increased dramatically.

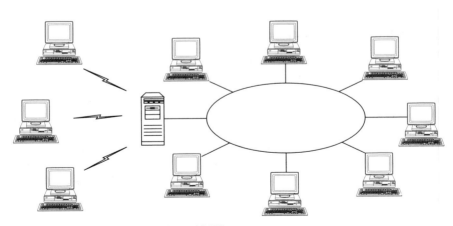

Figure 5.1 Integrated wireless and wired LANs.

Cellular phones wouldn't be nearly as useful if you could only talk to other cellular users. It's the fact that you can contact any other telephone user, cellular or land-line, that makes them attractive. And the fact that any other telephone user can call you. It's the combination of wired and wireless communications that provides benefits to both types of users. So you might see stand-alone wireless LANs, but only for certain applications. Expect to integrate any wireless LANs with existing wired networks.

Where wireless LANs make sense

Should you use a wireless LAN to supplement or replace a wired LAN? It depends on your circumstances. There are many bad reasons to go wireless, like trying to be the first person on your corporate block to have one, or making up for the fact that you never had a walkie-talkie as a kid. However, there are many good reasons for looking at wireless technology. In summary, wireless LANs make sense where:

- Conventional wiring cannot be used or can only be used with great difficulty.

- Conventional wiring could be used, but wireless will provide better results.

- Repeated temporary setups and relocations make wiring too awkward, costly, or time-consuming.

- Greater mobility is needed for LAN users.

We'll look at examples of each of these areas in the following sections.

Where wiring cannot be used

Depending on where your office or factory is located, wiring might not pose much of a problem. Many newer buildings include wiring ducts and closets that make the job of running cables quite straightforward. These arrangements are neat, safe (nothing to trip over), and can accommodate just about any type and grade of cable required, including fiber optic. If you are located in such a facility, count your blessings.

However, there are many situations where wiring cannot be used, or other cases where wiring is very difficult to do or the cables are vulnerable to damage. Wires can even pose a safety hazard. The following are some examples.

Historic or architecturally significant buildings. Buildings used for offices in many of the great cities of Europe might date back hundreds of years. Historic sites such as museums, legislatures, even palaces, all have need for computerized information systems, including LANs. But the last thing they want to do is start drilling holes or stringing cables along the walls. Coaxial cable running down the middle of the Louvre would probably not be well regarded.

The architectural features of some buildings, even newer ones, might also make running wires a difficult proposition. Consider the need to span an atrium rising up many floors. The cable run around the obstacle might involve a huge distance. Or consider the requirement to avoid running cables where they could detract from the interior design of the facility. Inconspicuous wireless nodes can connect computers without damaging buildings or detracting from their splendor.

Factory locations. PCs are becoming increasingly common on the shop floor, both for data collection and data delivery. Unfortunately, an active factory site can be difficult or impossible to wire. Floors formed from the thick concrete slabs necessary to support heavy machinery are not suitable for running wires through. Cabling along the ground poses a safety hazard to factory personnel and can be severed by equipment. Overhead cranes can rule out suspending wires from the ceiling. So conventional wiring might be out of the question. Rather than going without communications, a wireless LAN can connect shop PCs to the rest of the company's information systems. See Figure 5.2 for an illustration.

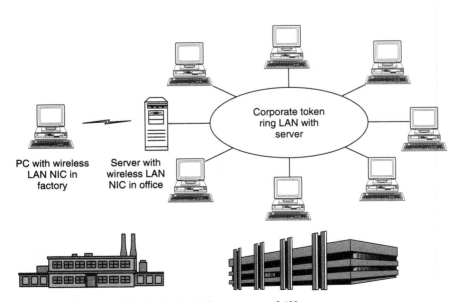

Figure 5.2 Shop floor PC with wireless link to corporate LAN.

Hazardous materials. Buildings containing hazardous materials, such as asbestos, in the ceilings and walls often cannot be wired in the normal way. To do so would disturb the material, creating a health hazard. While local building or health codes might require that the offending material be removed, this is usually an extremely expensive process undertaken by the landlord. The timing will be set by legal requirements, lease obligations, and cash flow. The need for LAN wiring probably won't figure quite as prominently. Until the hazard has been cleared, a wireless LAN can fill the need to interconnect computers.

Sealed rooms. Some areas of the office or factory cannot have just any old cable running in or out. For example, "clean rooms" used for research and for certain production processes are sealed to prevent dust or other contaminants from entering. Normal wiring conduits cannot be used without special protection. However, PCs used in the clean room can still have a requirement for connection to the LAN. A wireless LAN can penetrate building materials used for the clean room, such as glass and walls, without the need for cabling.

Where wireless is superior

In some cases, a wired link is perfectly feasible from a physical point of view. However, a wireless link might still be preferable because it offers a better combination of price and performance.

Connection of remote nodes. Imagine a scenario where one or two nodes of a LAN are in a remote location, say an area of the building that is isolated from the rest, or a second building on the same property. Providing a full-speed peer-to-peer connection to a wired LAN can be expensive because it involves running high-quality cables over long distances. Wiring costs might include trenching, poles, or other supports for the cable.

One solution is to link the nodes using a wireless radio LAN between the remote site and the main wired portion of the LAN. The remote nodes might not receive the same data transfer speeds, but they will still operate almost transparently as peers on the main LAN (see Figure 5.3).

Hostile environments. Some locations can cause problems for wired links. Equipment that uses high voltage like arc welders, or high current like electrical motors on a construction crane, can generate electrical noise that interferes with signals on a wired LAN. A wired link, while physically possible, might prove unusable.

In these cases, a wireless LAN might actually be more effective at carrying the data. The wireless technology features increased error checking for more reliable performance under adverse conditions. And since there is no

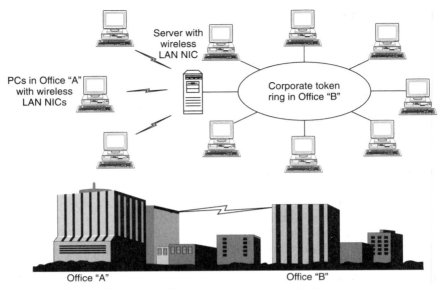

Figure 5.3 Remote PCs linked to office computers via wireless LAN.

cable to induce electrical noise in, the wireless LAN might be less susceptible to the interference being generated (particularly using a robust modulation technique such as direct sequence spread-spectrum).

High wiring cost. Using wires for some applications can end up being very expensive, even when it is possible to do. For example, in some types of factories, building and safety codes require that any wire needs to be encased in expensive fire-resistant conduits. For data communications around a huge outdoor recreational facility or a race track, the miles of wiring required could cost a small fortune. Also, consider a wire needed between two close buildings in a city. The connection might fall under the exclusive domain of a nation's Postal, Telephone and Telegraph (PTT) agency, which charges dearly for the link. The PTT might also be very slow to install the required line, and that delay carries with it an implicit opportunity cost. Or, in some countries where the telecommunications infrastructure is not well developed, the PTT might lack the ability to provide a reliable connection at all. In these cases, a wireless LAN can offer a far more cost-effective solution.

Temporary Locations

Sometimes, there is only a short-term need for a LAN at a given location. Examples include:

- Demonstration exhibits at trade shows.

- Work locations for teams like groups of auditors, who move from client to client.

- Offices used during political campaigns.

- Training courses given at different customer locations and conference facilities.

In some offices, staff relocations occur with enough frequency to justify considering a work location as a temporary site. Other offices, knowing that a move is coming soon, might resist installing a permanent LAN, instead of opting for temporary facilities.

A unique instance of a temporary LAN is used at a disaster recovery site. If a company has suffered a major disaster at their main office and computer facilities, they are forced to relocate to other premises. The relocation is normally temporary, until their original facilities are restored. So their need for a LAN at the recovery site is only for the duration of the disaster. You learn more about this in the following chapter.

When a LAN must be set up and torn down regularly, wiring can be awkward to reinstall and bulky to carry around. In this type of environment, wiring and connectors can become worn or damaged. Wireless LANs offer a very effective alternative to wires and cables when the LAN itself is constantly on the move.

Greater Mobility

Perhaps the best case for wireless LANs can be made when staff mobility is key to the application. In some cases, the application probably wouldn't be undertaken unless wireless technology enabled it. In others, wireless technology just makes life a lot easier.

The most obvious example relates to the newest selection of portable notebook and subnotebook computers, particularly those using PCMCIA cards. After cutting the power cord, and perhaps even the modem cable (by using cellular modem) it seems a shame to force the notebook users to wire themselves into the corporate LAN. Installing a wireless LAN network interface card in a portable computer allows LAN access throughout a selected area of an office or factory. Of course, the area is limited. They won't use the same means of connection from another city. However, it does provide freedom of movement between offices and meeting rooms, without having to reconnect to wired LAN outlets.

Wireless LANs can also make users more productive by freeing them from the need to return to wired terminals or computers. For example:

- Stock brokerage staff working on the trading floor can retrieve prices and record trades through wireless links.

- Retail stores can set up wireless point-of-sale registers anywhere in the store to enhance promotions of particular items.

- Health care professionals can review a patient's status and enter directions and treatment details using a wireless device directly at the patient's side.

- Warehouse staff can receive and record pick information on a computer directly from their forklifts or trucks.

- Restaurant staff can use wireless terminals to provide up-to-the-minute menu offerings and record customer selections at the table.

- Robots can receive direction and provide information collected without the need for stopping to transfer data.

As you can see, there are many ways a wireless LAN can be used to enhance business systems. These examples probably gave you some ideas on how you could use a wireless LAN for your applications. In the next two chapters, you'll see case studies of how others have used wireless LAN technology in their businesses.

The Applications

"Few things are harder to put up with than the annoyance of a good example."

Mark Twain

"Example is not the main thing in life—it is the only thing."

Albert Schweitzer

"The first great gift we can bestow on others is a good example."

Morell

6

Wireless Applications

So far, the material in this book has dealt with the technical aspects of wireless LAN technology. You now should have a good grasp of radio and light systems, including where they can be used. The purpose of this part of the book, starting with this chapter, is to introduce you to the promise of wireless. Knowing about the technology is not enough when you cannot imagine how it could be used. Hopefully, after you have studied some real-life applications, you can start to plan yours. This part of the book will not provide the technical specifications for each application, but rather the business use.

Some applications lend themselves to a particular technology (for example, microwave for high-speed, short-haul transmissions). This should not be construed as an endorsement of any solution or technology, but rather as a departure point for our discussion of wireless applications. The technology is evolving so fast that you will really need to investigate every possible solution before selecting yours. Further, one wireless technology might not satisfy your needs, and you might need to mix wired LAN components with radio and light components. You might have to have a very eclectic network to satisfy your needs.

Vendors are all too happy to sell you their product as the only solution. But you always should remember the Latin phrase "*caveat emptor*," buyer beware. The buyer should always be aware when buying nonstandard proprietary products. The LAN industry has seen this before. The wireless market is similar to the wired LAN market of the mid 1980s. There are lots of vendors with lots of proprietary products. Each of these vendors is jockeying for position with the standards committees. Until wireless vendors get serious about interoperability, and customers demand standards, you could purchase a solution today that is obsolete tomorrow. And I should know. I bought a SONY Beta videocassette recorder in 1978! On the other hand, the solutions you purchase today might provide a strong basis for all your future wireless communications.

Perhaps the best strategy for today, given the state of the technology, is to select applications that work and provide timely and adequate payback. The applications and their relative paybacks should drive your decision making.

Thinking of applications for wireless technologies is a good exercise in creative thinking. Fortunately for you, many applications already have been thought of and implemented by others. Later in the chapter, you will look at some innovative uses of wireless that do not fit in either the health care, financial, service, or retail industries (for these applications, see chapter 7).

Why would you want to consider using a wireless LAN? Besides your personal rewards, there are some other compelling reasons. Until recently, wired LANs were the only alternative providing users with connectivity in the rapidly evolving workplace. But the rapidly evolving workplace created the need for flexibility in workstation cabling. This need strains most existing cabling systems. The more conventional cabled LANs have certain problems associated with their cabling that wireless LANs propose to solve by simply doing away with cables. Specifically, the use of wireless communications will untether devices, suggesting fast and simple network installation and reconfiguration and opening the door to mobile computing. Table 6.1 reflects some issues you might want to consider when looking at whether to acquire and install a wired or wireless LAN. As you can see from Table 6.1, there are more considerations than just flexibility. You also will need to look at economic, legal, physical, and aesthetic issues, as well as flexibility.

Wireless communications, presently enjoying a boom in the mobile telephony market, are now entering the local area network (LAN) market. As the traditional medium of a LAN, cabling has certain drawbacks when seen in the context of the ever-changing scenery formed by an organization's demands for interconnecting computers. First, installing new cabling can be expensive and tedious, and second, changing the cabling involves more of the same costs and effort while possibly (temporarily) disrupting the operation in parts of the network. This is especially true in the case of the older, rugged cabling varieties. Newer structured cabling systems use suppler unshielded twisted-pair cables and incorporate redundancy to lessen these drawbacks. Nevertheless, the drawbacks remain significant in situations demanding frequent changes to a LAN.

Finally, wireless communications offer an aesthetic solution to these problems by doing away with all the cables. A wireless LAN thus suggests less expensive, fast, and simple network installment and reconfiguration. What is more, a wireless LAN opens the door to mobile computing. Wireless communications untether devices and give them continuous access to the network anywhere within the local area.

TABLE 6.1   Wireless Versus Wired Considerations

Issue	Wireless	Wired
Costs	Acquisition costs high, maintenance costs low.	Acquisition costs low, maintenance costs high.
Flexibility	Workstations can move or join network easily.	Workstations fixed in location.
Legal	May require licensing, generally spread-spectrum does not.	Most buildings require the installation of special plenum cables to meet fire regulations. May be legal requirements regarding the installation of cable where there is asbestos in the walls or ceiling.
Physical	Light will not travel through walls.	Requires conduit space and easy access to floors. Coring of floors may be required, and building owner may not approve.
Aesthetics	Most devices are aesthetically designed to fit in to the office environment.	Walls and power cables are accessible and built-in and wiring will not affect the look of the office.

Generally, wireless LANs provide:

- Flexibility. They create an untethered environment for the user.
- Cost-effectiveness. They are less expensive in the long-run for dynamic environments.

On the other hand, wireless LANs are:

- Slower. They generally have lower transmission rates than wired LANs.
- More expensive to install initially.
- Susceptible to interference or limited in distance, depending on the technology.

Having said that, wireless LANs theoretically form a viable proposition as a networking solution in four broad functional categories: flexible networks, temporary networks, networks with mobile devices, and networks in environments where cabling is difficult. The use of a wireless LAN does, however, have several (negative) consequences (most notably for network security) unless the products fulfill certain requirements.

The current generation of wireless LAN products weakens the previous proposition. Some products are only suitable as a flexible network in open workspaces, and only some products are well suited to networks with mobile devices. The requirements are only partially satisfied.

Subsequently, the current wireless LAN products are only a suitable contender as a partial networking solution. That is to say that the products can fill niche applications. Nevertheless, these niches can be found throughout the gamut of LAN applications.

Wireless LANs need not remain niche solutions in the case of flexible networks and networks with mobile devices. However, use as a flexible network faces stiff competition from structured cabling, especially if capacity demands are high. Standards are evolving for wired 100-Mbps Ethernet and Token Ring. Future products are expected to have an increased capacity, but this will not always be of help because the user demands for capacity will likewise increase. Client/sever computing applications proposed by some organizations definitely will require more than 2-Mbps data rates.

Growth in portable applications is thus seen as a prerequisite to the achievement of a significant market share for wireless LANs. The size of this share also will determine whether future products will satisfy more of the formulated requirements. Now that you have reviewed the requirements for wireless, you are ready to review the applications.

Industrial Applications

As a result of growth in computer-controlled manufacturing, industrial applications are a large wireless LAN marketplace. Products and other equipment manufactured throughout the world are increasingly being designed to interface with computers, especially personal computers. Wireless LAN technology is particularly suited to provide the linkage between such products and equipment to the computers. There are many instances where wireless technology could be used in industrial applications to link computers and equipment.

At least four converging demands for wireless LANs spring to mind as potential applications. First, the programmable logic controllers (PLCs) for controlling automated equipment currently form very sophisticated control systems for programming using computers. Second, process automation is continually moving to larger integrated systems, using a local area network to interconnect decentralized PLCs and any autonomous equipment. Third, the advance of computer-aided systems from design (CAD) to manufacturing and engineering (CAM/CAE) created the demand for the integration of high-end workstations with the LAN. And last, the large pool of mobile devices and users for production and transportation, such as mobile robots and automatic guided vehicles (AGVs), will increase sharply with the growth in just-in-time manufacturing techniques.

Industrial applications call for extremely reliable networks that can accommodate a wide range of fixed and mobile devices. The coverage area can be expansive, usually consisting of several buildings. The capacity demands, however, are moderate, with the possible exception of CAD/CAM/CAE workstations.

Some industrial environments prohibit, restrict, or make extremely difficult the use of cabling (for example, where the building, factory, or plant has massive concrete floors and high bay ceilings, or where the cabling must enter either clean or hermetically sealed rooms). Other industrial environments call for thorough shielding of the cables against very high levels of interference from machinery or other radio-controlled devices present in most industrial settings.

Other considerations sometimes dictate the use of wireless networks. For instance, in the large construction halls in the aircraft industry, the assembly line is built around the aircraft rather than the other way around. There are other instances where networks are needed for various short-term projects or jobs. Let's look at some of the excellent examples of wireless LANs in industrial settings.

Petroleum industry

The oil patch is rough terrain not normally conducive to the laying of cable. So wireless technology can find a niche in this environment. Wireless can provide a data link from an oil-drilling rig to an on-site contractor's shed for the purposes of displaying and logging important geological data obtained during the drilling operation.Also, stringing wires through an oil refinery might pose an electrical hazard, so wireless LANs often connect workstations.

Agriculture and food

A farmer can use a data link from a coding device to a computer in dairy farming applications. The coding device reads an animal's identification number that is magnetically read from a small apparatus inserted under the skin of a milking cow. The computer keeps track of the cow's milk output and important information concerning the health of the animal.

Factory control

Factories are a prime candidate for wireless. There are several reasons for this. On factory floors, you regularly find that running a cable to personal computers is unworkable. Often, you cannot run cable under the floor because it has no basement and is made from reinforced concrete to support heavy machinery. Furthermore, overhead cranes make overhead runs difficult. Besides, you usually cannot lay cable on the floor because of forklifts and heavy dollies moving about the factory. By the time you figure out how to string cable to the personal computer, they decide to move the workstation and PC. For these reasons and more, wireless is an attractive option for factories.

There are many potential factory control applications. Wireless technology can provide a data link in the following applications:

- Data acquisition equipment and a personal computer.
- Engine compartment of an automobile to a diagnostics computer.
- Industrial equipment in a plastics factory.
- Inventory control.
- Numerically controlled metal working equipment.
- Overhead crane control.
- Portable weighing scales to a computer keeping records.
- Robotics.
- Scoreboards.

Data acquisition equipment and a personal computer. More than one organization is linking data acquisition equipment to a personal computer. The data acquisition equipment produces digital representations of waveforms. The personal computer displays these waveforms and performs various signal analyses on the waveforms. By using a wireless link, the data acquisition equipment could be moved within the facility without the need to move the heavier and larger computer. This set up, with the wireless link, can provide a low-cost alternative to an expensive digital signal processing oscilloscope, which is traditionally used for waveform acquisition and signal analysis. Figure 6.1 provides an example of a wireless link to industrial, scientific, or medical equipment.

Engine compartment of an automobile to a diagnostics computer. Wireless technology can provide a data link from the engine compartment of an automobile to a diagnostics computer within the cab of the automobile so that vehicle emissions could be monitored as the automobile is driven. In one case, an engineer for a company that manufactured and sold vehicle diagnostics equipment claimed that a moving road test was the only reliable approach to testing any vehicle's emissions. A wireless link provides a convenient means

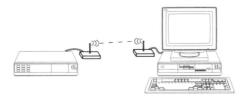

Figure 6.1 Wireless link to industrial, scientific, or medical equipment.

for the diagnostics computer to link to the vehicle's electronics test port. Most automobiles presently being manufactured include a serial port that is compatible with a wireless data link. This serial port, which is normally within the engine compartment, is used to program and read data to the automobile electronics processing unit.

Industrial equipment in a plastics factory. Any wiring in a plastics factory, or any other facility that uses petroleum in its process (for example, most chemical plants, refineries, and plastics manufacturing) must be enclosed within an expensive fire-resistant conduit. Therefore, in environments such as this, wireless is a much lower cost alternative for the data link.

Inventory control. To aid the process of shipping and receiving of parts and inventory, a wireless link can be used to provide a wireless interface directly with bar-code readers, notebook computers, and a central data-processing computer.

Numerically controlled metalworking equipment. A wireless link provides a very convenient approach to downloading machine operation codes into the metalworking equipment's control unit. The environments in the immediate vicinity of metalworking equipment are not hospitable to computers and computer programmers due to noise, vibration, metal shavings, and cutting oils. Therefore, there is a demand for convenient approaches to linking computers to the numerically controlled equipment where the computer is separated at some distance or in a separate room.

Overhead crane control. A wireless data link can replace another type of data linkage product that had been used previously by the end user. The previous approach to data linkage was by means of a 200-kHz signal transmitted through the power lines (usually referred to as "carrier current"). A 900-MHz radio signal link can prove to be much more reliable than the 200-kHz power-line approach. The "carrier current" power-line data link was easily jammed by motors and other industrial equipment within the same facility, whereas the 900-MHz radio link is not disturbed by these same interfering factors.

Portable weighing scales to a computer keeping records. A wireless link can be used with portable weighing scales to send data from the scales to a computer that keeps records of the measured weights and other data. The wireless link enhances the convenience of a portable scale, especially in facilities where the scales are constantly being moved to weigh heavy materials, boxes, and shipping containers.

Robotics. A research laboratory associated with a major university is using a wireless data link to provide computer-controlled movements of a small robot.

The controlling computer is bidirectionally and wirelessly linked to the robot, with sensor data being sent from the robot to the computer while, simultaneously, motion commands are being sent from the computer to the robot.

Scoreboards. Wireless data links are being used to send data to sports scoreboards located in field houses, stadiums, and ballparks. Modern sports scoreboards being installed in schools, parks, and arenas are normally interfaced with a control unit (or a personal computer being used as a control unit) through a serial port. The use of a wireless link to the serial port provides cost savings over the traditional hardwired methods that normally require expensive conduit installations. Furthermore, portability of the control unit (or personal computer), as provided by a wireless link, is a desirable feature, especially at facilities that are multipurpose and provide support for many sporting games, such as baseball and football.

In short, networks for industrial applications cover the whole range of wireless LAN functionality, with special emphasis on networks with mobile devices. No one wireless LAN product can satisfy all demands. For instance, the desire to connect high-end workstations to the LAN and the need to cover a wide area will make the use of cables in at least part of the network inevitable.

Transportation Industry

Organizations operating a large fleet of vehicles, vessels, or aircraft are traditional users of wide area networks for mobile voice communication. But these same organizations are adding mobile data communications at an alarming rate. As a result of a move towards fleet management systems, these organizations also are developing a need for local and wide area data networking. An increasing number of vehicles have a terminal informing the driver of the route and information about their pickups and drop-offs, using wide area networks that are perhaps satellite-based. But there also are wireless local area network applications where the fleet management functions call for intermittent connections. These functions might include preventive vehicle maintenance, where on-board data acquisition devices can collect data on the vehicle's condition during trips and upload it on the vehicle's return.

Organizations in the transportation industry have found innovative ways to use wireless technology to provide a data link as follows:

- A ship and oceanographic research buoys.
- Programmable road signs.
- A "scaled" helicopter and a flight control computer.
- Devices measuring speed and weight of vehicles traveling on roads and highways.

A ship and oceanographic research buoys. Wireless technology is used to gather data from buoys. The buoys include sensors and electronics to monitor and store data representing important ecological and meteorological information. Wireless can provide a means for transferring this data to a passing research vessel. Wireless provides a more convenient alternative to the normal method of maneuvering a ship next to a buoy and actually connecting a cable between the ship and the buoy. The buoy might not be in the calmest ocean waters which makes the wired cable link difficult and possibly hazardous to connect between the ship and buoy. Perhaps this application stretches your definition of a local area network!

Programmable road signs. Signs can be either temporarily installed to display information during road construction or special events, or permanently installed to show current road conditions. Wireless can replace the current method of programming the signs through cellular phone radio links.

A "scaled" helicopter and a flight control computer. A helicopter is being flown by remote computer control without any human intervention. This "computer-controlled autopilot" system has been developed by a major university as an entry into a national competitive event between engineering schools.

Devices measuring speed and weight of vehicles traveling on roads and highways. Governments are looking at wireless as a means to measure the speed and weight of vehicles traveling on roads and highways and to transmit the data to a computer located within a law enforcement vehicle. The speed/weight measurement device is located beside the road and connects to thin strip sensors draped across the road. A wireless link allows the law enforcement vehicle to either acquire the speed/weight data while moving or to monitor speed and weight data while at a less conspicuous location. So now truckers will have more to worry about than radar and photo-radar! Using a wireless LAN instead of a wireless WAN can be advantageous when the costs of the LAN, a private network, are less than the fees paid to the carrier in the public WAN.

Office and Similar Applications

Figure 6.2 shows how wireless local area networks fit into today's office applications and environment. The diagram is divided into 4 quadrants, labeled from 1 to 4. The X-axis goes from no mobility at the bottom to total mobility at the top. The Y-axis goes from local applications on the left to wide area or remote applications at the far right.

Starting in the lower left-hand corner, you will find quadrant 1. Quadrant 1 is where many companies are today. You see desktop computers that are

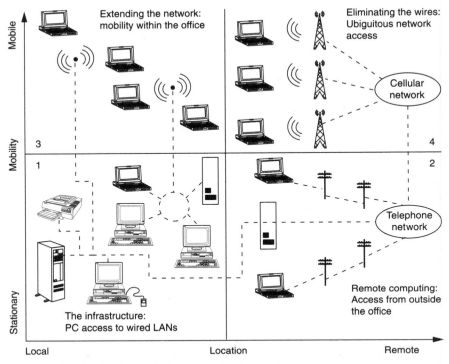

Figure 6.2 Wireless LANs in today's office applications and environment.

hardwired to either an Ethernet or Token Ring LAN. Using an adapter card, you can have notebook or palmtop computer users connecting to cabling in any office or cubicle. As consultants, we often connect in this manner, using our notebook computers with Ethernet adapters that allow us to connect to a client's network to use electronic mail or scheduling. Several vendors offer pocket adapters that can support Token Ring or Ethernet, or shielded-twisted-pair, thin-net or thick-net wiring topologies. This type of network access can be classified as stationary and local. You still must come in to the office and find an unused connection point to access the network.

Moving to the right in Figure 6.2, you will find quadrant 2. In this quadrant, you again have a familiar scenario for most organizations. Maybe your organization has a remote office that accesses the network. Because their traffic volume is low, they cannot justify the cost of a leased line. Instead, these remote sites dial in to the network using either the plain old telephone system (that is, POTS) or a packet-switched network (for example, Tymnet). Perhaps your organization also has a network administrator who can dial in to the network and use either NetWare's REMOTE, Carbon Copy, or PC Anywhere. These products allow the user to remotely control a PC or server. There are many instances of network access using the wired telephone system. Again, this type of network access can be classified as stationary and remote. These access points are remote, but they are stationary in that they must somehow connect to an access point on the wired network, through, for instance, a modem.

If you look upward from quadrant 1 in Figure 6.2, you will find quadrant 3. Quadrant 3 is the basis for this book—wireless LANs. With wireless LANs, you have total mobility within your office. You are free to roam about your workplace, moving from your office to a conference room or even to another building on a campus, while remaining seamlessly connected to the network. This arrangement is classified as local, remote computing.

Finally, in quadrant 4 you will find users with total freedom, who can access the network wherever there is cellular service offered. From chapter 1, recall that these networks are metropolitan or wide area networks. Access to the network through a MAN or WAN can be classified as remote and mobile. Examples of metropolitan or wide area networking using cellular technology abound. Exploding on to the scene are paging and electronic mail systems using cellular technology. What might not be so obvious are the uses of spread-spectrum, infrared, and microwave technology for local area office applications.

So far, our examples have been primarily for industrial or transportation applications. Similarly, wireless can provide a data link in the following office or nonindustrial applications:

- Braille display units.
- A personal computer to a wired computer network.
- Printer sharing.
- Remote control.
- Terminals and computers in a courthouse.
- Electronic mail.
- Office sharing.

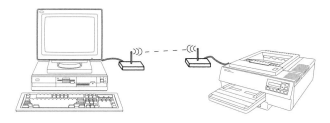

Figure 6.3 Wireless computer-to-peripheral link.

Braille display units. Wireless can provide a means of linking computers that display in Braille for teaching blind and deaf-blind people. A manufacturer of Braille computing equipment is investigating using wireless as a means to link the instructor's computer to the student's computer during class. A means of linkage is most important for instruction to the deaf-blind. Furthermore, a wireless link renders portability so that the Braille computer can be easily used in many classrooms and at home.

A personal computer to a wired computer network. Wireless is being used to access a network from a lone computer in a separate building, across a parking lot, across a road, or just in an inconvenient location. The traditional alternatives to wireless is to install an underground conduit for cabling to the network, or when the lone computer is on the other side of a public road, to lease a dedicated phone line. These alternatives are much more expensive than a wireless data link.

Printer sharing. Wireless is not only being used for wireless printer sharing, but also for linking to a distant printer within a shipping department where shipping orders are being printed. You can create a wireless connection between your computer and any printer that has a transceiver attached. Figure 6.3 provides an example of a computer with a wireless link to a printer. This wireless connection allows you to position the printer wherever it's most convenient for you. You're no longer bound by the 10-foot-length limit of standard parallel printer cables, and you don't have to worry about rearranging cables if you decide to move the printer. Also, you can share a local printer that's attached to a computer with a wireless transceiver, as shown in Figure 6.4.

Remote control. You can use wireless to establish a link between an executive's desktop computer and her secretary's computer to allow printer and file sharing. Figure 6.5 shows an example of an executive's portable linking to the secretary's desktop.

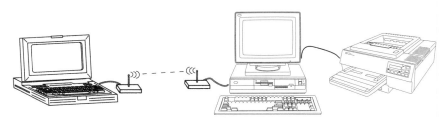

Figure 6.4 Wireless link to a peripheral locally attached to a computer.

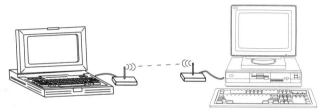

Figure 6.5 Wireless computer-to-computer link.

Terminals and computers in a courthouse. Terminals and computers are being used for acquisition of legal case history during trials and hearings. A wireless data link provides a low-cost approach to providing a courthouse network with software, hardware, and installations for courthouse network applications. These also can be used in hospitals, where wireless might be especially important for linking a doctor's or nurse's pen-based portable computer to the hospital's central computer for acquiring and updating medical records.

Electronic mail. Wireless LANs are good for electronic mail. By using public electronic mail gateways and wireless LANs, you can send and receive your e-mail from anywhere. While wireless e-mail today represents a very small percentage of the messaging market, chances are it will experience phenomenal growth. Today, there are 15 million users of public and private electronic mail systems, but most of them are connected by local area network cabling or the plain old telephone system. As you can see from Figure 6.6, the Electronic Mail Association estimates that by 1995 that there will be approximately 27 million e-mail users. A growing percentage of these electronic mail users will be doing their business outside of their office.

Office sharing. In major urban centers, office space is at a premium in the downtown core. Consulting firms locate in the downtown core to be near their clients. Understandably, this office space is more expensive than real estate in less desirable areas. Because of these higher costs, many consulting firms are asking consultants to share workspaces. They schedule the use of the workspace, then come in at the scheduled time and access the network through a wireless connection. This means different consultants do not need to share a single computer.

As you can see or imagine, there are many wireless LAN applications because LANs are predominantly used in office settings. Originally used to connect desktop computers and printers, there is an abundance of wired LAN solutions. It is unlikely that they will be replaced in the short term. The capacity of current wireless LAN products will not satisfy all users' demands, but wireless products will find a niche where organizations cannot

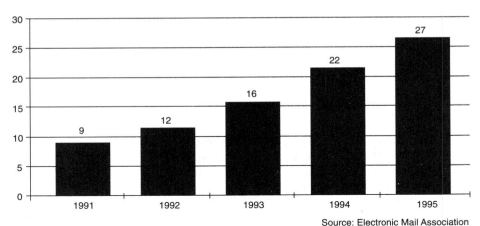

Figure 6.6 Electronic mail users.

economically lay cable, or where there is a need to extend a network, or where a temporary network is needed.

Wireless Disaster Recovery Applications

If ever there was an application made for wireless local area networking, it is disaster recovery. Wireless offers flexibility and mobility that's unsurpassed in the wired world, attributes that are desirable when working through a disaster. Many organizations started using wireless networks for backup and recovery and then ended up using wireless as the primary cabling media.

How can wireless help in a disaster? Well, it's not hard to imagine how cellular communications could be used to recover your voice network. You could use a cellular link to route your calls to the outside world in the event of the loss of your telephone trunk. If you lost your private branch exchange (PBX), then you could easily install a wireless PBX. Various vendors are offering a PBX "on wheels" that could be brought to your site as a short-term wireless solution.

Harder to imagine is how you could use wireless LAN technology when facing a disaster situation. Many organizations have used microwave technology to provide large amounts of bandwidth in the T1/E1 range to their central office (CO) or to their interchange carrier's (IXC) points of presence (POP). You even could use microwave to backup your leased lines in your campus area network.

A more likely scenario actually involves wireless LAN technology itself. Suppose you have wiring damaged by water during a flood or burnt during a fire. Normally, you would need to rewire the LAN. However, if you had wireless LAN technology as backup, you would just bring the new equipment in and be back up in a matter of hours, not days. Of course, if the building as well as the wiring is gone, a wireless LAN is the quickest way to get back up and running at a new location. Wireless LAN technology has been used in several serious calamities.

During recovery mode, you can link your LAN and PBX together using an infrared laser, an unlicensed technology providing short-haul, relatively high-speed transmissions. Table 6.2 gives some insight into the use of wireless LAN technology during disaster situations.

Wireless Considerations

When designing your wireless network, consider designing it like you would a wide area network. In a wide area network, you usually locate large programs and data files locally. The same considerations should be given to your wireless network. Loading programs over a wireless LAN at the slower speeds could prove most frustrating to most users. Of course, you can alleviate some wireless LAN performance problems by designing your applications so only small parts of your program load as necessary. You should not rely on most applications to be aware of the bandwidth restrictions of your wireless network.

TABLE 6.2 Wireless Disaster Recovery Instances

Location	Disaster	Comments
Chicago, IL	Flood	Wireless data and LANs
Desert Storm, Iraq	War	Microwave and wireless LANs and PBXs
Hinsdale, IL	Fire	Infrared
New York, NY	Bombing (WTC)	Microwave and wireless LANs
San Francisco, CA	Earthquakes	Microwave and spread-spectrum wireless LANs
South/North Carolina	Hurricane Hugo	Microwave used extensively
Southeast Florida	Hurricane Andrew	Spread-spectrum and infrared wireless LANs

As you can see in this chapter, there are many good applications for wireless networking. So, let's look at some of the criteria you should look at when considering the use of wireless for your application. Table 6.3 provides a checklist for assessing the suitability of wireless. If you cannot answer "yes" to at least 50 percent of the questions, then you should revisit your decision to install a wireless LAN. You would probably be better off on a wired LAN. If nothing else, answering the simple checklist might cause you to reflect on your decision and help you decide a likely direction.

If you still have not made up your mind, you might want to look at the specific examples provided in chapter 7 (for the health care, financial, service, and retail industries). It's now time to turn our attention to some very interesting applications of wireless in the health care industry.

TABLE 6.3 Wireless Selection Checklist

Question	Yes/No
The Selection Process	
1. The existing wired network cannot be used.	
2. You want to prototype the technology.	
3. You have looked at all other options.	
4. Other companies are successfully using the technology.	
5. The long-term costs for the technology are less than a wired solution.	
6. The application could not be offered in a wired environment.	
7. The selected technology is the best solution among the wireless alternatives.	
Technology Issues	
1. The physical and electrical requirements for this technology are understood.	
2. The rated throughput is adequate for the application.	
3. The office cannot be wired because of asbestos.	
4. The network is temporary.	
Standards Environment	
1. There are de jure standards for this offering.	
2. If de jure standards do not exist, there are de facto standards or generally-accepted implementations for the technology.	
Administrative Management Issues	
1. Licensing is not an issue.	
2. There are no legal issues regarding the technology or the proposed solution.	
3. Security is not an issue.	
4. The health and safety of employees is not compromised.	
Network Management Issues	
1. Standard network management software is provided with the technology.	
2. The hardware can be remotely monitored.	

7

Industry Applications

The last chapter looked at wireless applications in general and at various industrial applications in specific. By now, you should be thinking of ways you could implement wireless technology in your organization. This chapter introduces you to more applications from the health care, financial, service, and retail industries.

Health Care Industry

Health care professionals are calling for access to patient information on demand. When doctors enter a patient's room, they want access to current information regarding blood tests taken that morning. They also want a full medical history for the patient before prescribing any medication or therapy. Accordingly, health care system designers are looking at wireless. The health care industry computer network designers are choosing spread-spectrum and infrared wireless local area networks as viable solutions where system requirements call for a high degree of mobility and installation in hard-to-wire places. The proliferation of portable computers, personal digital assistants, hand-held communicators, and the availability of wireless interfaces fitting into a portable computer's PCMCIA (Personal Computer Memory Card International Association) interface slot is driving the steady growth and sales of wireless products. Health care is one industry benefiting greatly from the use of portable computing devices and wireless local area network technology.

Most hospitals have a large administrative system, numerous patient monitoring devices, computer-controlled medical equipment, and a supplies inventory system. There is a latent demand by the health care professionals for access to these information sources using hand-held communicators from anywhere in the hospital. Such a communicator would allow a doctor or nurse to access the patient's medical record bedside without having to carry heavy hard-copy files, to access a centralized expert system to aid diagnosis or predict the interaction and reaction of different drugs, to call other doctors or specialists for assistance, to send urgent messages to staff, and to enter current data on the patient's condition.

Health care wireless techniques

Wireless networks provide connectivity between computers by use of radio waves or infrared light. The use of radio waves is currently the most developed and popular method for providing wireless links between computers. Radio-based LANs use radio wave transceivers in broadcast mode to send data from one computer to another. In addition, some wireless implementations use modulated infrared light reflected off the ceiling to carry data between computers.

One vendor offers a personal assistant for nurses and doctors. This personal assistant puts the power of computers and communications technology in the hands of the health care professional. The device is a portable hand-held wireless data communicator that features a very easy-to-use, menu-driven touch-screen interface. The wireless communicator allows users to be online with the hospital's local area network.

For example, nurses equipped with the communicator start their shift by touching the screen for their assigned patients and room numbers. Touching the first patient's name will call up the prescribed care for that patient by radio link from the floor nurse's station. Prompts will appear on the screen. Any type of patient service (for example: blood pressure, temperature, and medication dosages) can be entered or displayed with a simple touch to the communicator's screen. By an RF radio link, the system can update and store all the patient's records and billing information.

You have seen repeated references to hand-held communicators. What does this hand-held communicator that health care professionals use look like? Well, in this book, hand-held communicators refer to any device that can be held in the hand, so it can include palmtop computers, personal digital assistants, or other specialized devices. These hand-held communicators also have a means for transmitting data, whether it be infrared or spread-spectrum.

HP markets palmtop computers (the HP 100LX or the HP Omnibook 530), as does Casio, Sharp, and other manufacturers. Apple, Sharp and Motorola manufacture personal digital assistants (PDAs). Table 7.1 shows a list of seven PDAs and their characteristics. Other specialized devices are manufactured by Psion, Granite, Dauphin, Fujitsu, et cetera. Many of these

TABLE 7.1 Examples of Personal Digital Assistants (PDAs)

Manufacturer	Apple	Apple	AST	Casio	Motorola	Sharp	Tandy
Model	Newton MessagePad 100	Newton MessagePad 110	GridPad 2390	Z-7000	Envoy	PI-7000 Expert	Z-PDA ZOOMER
Operating environment	Newton Intelligence	Newton Intelligence	GEOS, PenRight	GEOS	Magic Cap	Newton Intelligence	GEOS
Communications	Infrared	Infrared	Infrared	Infrared	Infrared	Infrared	Infrared
Desktop connection	PC Windows/Mac	PC Windows/Mac	PC	PC	PC	PC Windows/Mac	PC
Suggested price (US$)	$499	$599	$549	$599	$1,599	$899	$699

devices are pen-based. Table 7.2 provides a comparison of pen-based devices. These vendors recognize that hardware is the easy part, and they either develop or work with integrators to develop applications for the health care, financial, and service industries.

Now let's get back to our health care examples. In another health care illustration, a university hospital wanted to eliminate virtually all communications wiring at its six-building hospital and medical center complex.

The hospital chose hand-held communicators using spread-spectrum technology and desktop workstations with wireless network adapters. The hand-held communicator is a pen-based microcomputer that runs DOS and can transmit real-time data to the host system. These devices have terminal emulation software that enables the user to communicate without wires to the hospital health information system, the main administrative system. All functions available on a desktop terminal that's hardwired to the same system can be accessed on the hand-held communicator on its monochrome screen.

By October, 1993, the medical/surgical unit had 5 hand-held communicators. Doctors and nurses use the devices to access patients' clinical records, place orders, do charting, and perform other clinical functions. Shortly after this introduction, the hospital added another 25 devices for use throughout the premises.

Communication with the host system is through spread-spectrum routers/repeaters, small units mounted unobtrusively on the ceiling throughout. These devices form the wireless link between the hand-held communicators and the host system. The design calls for about three routers/repeaters per nursing floor.

The hospital also used the same technology at their cancer center, housed in a nearby building. There, wireless adapter cards were installed in desktop personal computers, which allowed these PCs to communicate with the host system.

As a side benefit, wireless PCs, normally used for backroom billing, can be brought quickly to the front office and used for admissions when the office is busy. Flexibility such as this is a major benefit of wireless and saves money when you move PCs.

The hospital also went wireless at the physical therapy building. There, personal computers were fitted with wireless adapters for admissions, clinical order entry, and information retrieval. These PCs also link with a minicomputer used for digital transcription. In this environment, users perceive no difference in response compared to their old wired system.

Because the hospital had such success using wireless intrabuilding, they decided to go wireless interbuilding. They installed a link between buildings two blocks apart to act as a backup to their fiber-optic cable.

TABLE 7.2 Examples of Pen-Based Systems

Manufacturer	Model	Operating environment	Communications	Desktop connection	Suggested price (US$)
AST	PalmPad SL	PenRight!, WinPen	Spread-spectrum	PC Windows/Mac	N/A
COMPAQ	Concerto 4/25 and 4/33	WinPen	Spread-spectrum	LAN	$1,299 (25 MHz) and $1,399 (33 MHz)
Dauphin	DTR-1	WinPen	Spread-spectrum	LAN	$2,495
Fujitsu	325Point	WinPen, PenPoint, PenDOS	Infrared	LAN card	$1,695
Fujitsu	325Point RF	WinPen, PenPoint, PenDOS	Spread-spectrum, infrared	LAN card	$2,530
Fujitsu	Poqetpad Plus	MS-DOS 5.0	Infrared	LAN, docking station	$1,450
Fujitsu	Pad Plus RF	MS-DOS 5.0	Spread-spectrum, infrared	LAN, docking station	$2,285
Kalidor	K2000	Windows, PenDOS, PenRight!	Infrared	File transfer and network utilities	$3,495
Norand	Pen*Key	WinPen, Power Pen Pal	Spread-spectrum	LAN	$3,195
Symbol	PPT4100	DR-DOS 5.0, PenDOS	Infrared	LAN	$3,895
Telepad SL	SL	WinPen, PenPoint, PenRight!	Spread-spectrum	PC	$2,495–2,795
Telepad	TelePad 3	WinPen, OS/2	Infrared	LAN, WAN	$2,995–5,995
Telxon	PTC-1140	MS-DOS 5.0, PenRight!	Spread-spectrum	N/A	$2,595
Telxon	PTC-1180	MS-DOS 5.0, PenRight!	Spread-spectrum	N/A	$2,595

System specialists are planning many other applications of wireless. The goal is to eliminate wiring and replace it with wireless wherever possible. At first, the wireless LAN equipment experienced minor problems due to some interference from hospital equipment, which they overcame by repositioning the repeaters/routers.

Wireless benefits in health care

People are constantly on the go in a hospital. Often it is difficult to install network wiring in hospital facilities. The cable-free environment of a wireless LAN provides many benefits for health care. By providing wireless portable computers to doctors, nurses, and hospital administration staff, a hospital can provide more timely and accurate care to patients. Some of the advantages of wireless computing are:

- Access to patient records and treatment information from practically anywhere.
- Complete prescription and medicine tracking from the patient's bedside.
- Location-independent claims processing.
- The extension of health care to the home.

The use of a hand-held device also keeps the nurse focused on patient care, resulting in faster, more accurate, and improved care. You can eliminate distracting paperwork and record keeping. Yet because you have moved the capture of data closer to the service by having the doctor or nurse enter it, billing can be faster and more accurate. And because the exact care is documented bedside, liability can be significantly reduced. Finally, touch-screen technology is extremely easy to use, so training time is reduced.

Wireless issues in health care

The main issues of using wireless networks to satisfy health care networking requirements are possible frequency interference, irregular propagation patterns, and the threat of someone stealing sensitive information. Wireless networking devices use frequencies similar to other wireless devices found in hospitals. Therefore, designers must be careful to specify a system that avoids interference with existing electronics. Hospital construction can highly distort radio wave propagation, which offers an installation hurdle that requires prudent planning for proper signal coverage. There is a requirement for support of many portable devices whose aggregate capacity demands are high and that provide full coverage throughout various buildings. Finally, be certain to provide proper security, such as encryption, when sending information over hospital wireless networks.

Financial Services Industry

Another industry testing the wireless airwaves is the financial industry. As businesses in the financial industry move to a marketing, customer-service orientation, you will see the spreading of wireless across the industry spectrum. Some of the early uses of wireless LANs have come in the financial markets and exchanges (for instance, the Commodity Exchange, the American Stock Exchange, and the Chicago Mercantile Exchange).

The Commodity Exchange, Inc. (COMEX) deals in futures and options for gold, silver, platinum, palladium, copper, and European equities. The exchange has implemented hand-held communicators on the trading-room floor to increase the speed, accuracy, and reliability of its price reporting system (PRS). They now have nearly 50 trading-floor personnel (called "price reporters") equipped with hand-held communicators that send data over a wireless local area network. This application is enabled by the integration of touch-screen technology with spread-spectrum radio transmission to allow COMEX employees to enter the information and automatically send it to the PRS through a wireless LAN.

The COMEX wireless LAN application was developed by Granite Communications. Granite integrates communications, microprocessing, software, and display features into hand-held personal data communicators connected by the Granite LINKS wireless networking software protocol. Granite's products also use VIDEOPAD, a touch-screen interface that replaces the conventional pen-and-paper pad used by traders.

Each communicator fits in the palm of the hand and weighs less than 20 ounces, so COMEX employees on the trading floor can enter the information easily, even in a space-constrained setting. The trading-ring reporters are prompted by the communicator and can enter information using the touch-screen. Virtually no training is required.

When a series of "buys" and "sells" results in a change in price for a given commodity, COMEX must report that price change to the market on a timely basis. In the past, price reporters would listen for a price change and use hand signals to alert a trading-floor supervisor, who would then call the price change through a headset to a data entry clerk away from the trading floor. The clerk would then enter the price into a minicomputer that would transfer the information across a high-speed ticker network to quote services and the wall display boards.

Now the price reporters use these hand-held communicators to transmit data at 56 kilobits per second to two antenna on the ceiling of the trading area. The two antennas are hard-wired to two personal computers in two different communications rooms. These PCs relay the data via a packet-switched network to two minicomputers used by COMEX's Commodity Exchange Center to send data to quote services such as Reuters.

The American Stock Exchange undertook a similar experiment. Some brokers and traders roamed the exchange's floor equipped with hand-held communicators that enabled them to get current price quotes and to execute trades.

Financial services wireless techniques

In the case of COMEX, trading-ring reporters now send real-time price changes to the PRS, which displays the prices on trading-floor screens and distributes the prices to traders through the market data services. The communicators send the data to the PRS over a wireless, spread-spectrum radio frequency LAN.

The hand-held devices communicate to a base station located on the ceiling of the trading floor. It transmits and receives data at data rates up to 121 Kbps and provides an RS-422 link to the host computer. The COMEX PRS using Granite's VIDEOPAD Links provides instantaneous transmittal of price information into the COMES computer systems, through a combination of front-end PCs, radio frequency products, and application software.

A personal computer converts the data into the proper format for the host system or hand-held remote, depending on the intended direction of the data. The host system is connected to the PC and uses Granite's Links protocol. The base station controllers are full-slot PC cards that provide the application program interface (API) and protocol management between the PC applications and the base station. Using a high-speed RS-422 data line, the base station communicates data to a base station. Figure 7.1 shows what this network looks like.

Wireless benefits in the financial services industry

COMEX has expanded the number of staff and is saving about $200,000 per year in data entry salaries, while improving the validity of the information and increasing the efficiency of the price reporting system.

For the American Stock Exchange brokers and traders, the communicators could hold information about 1,000 trades. This would be of benefit to the brokers and traders. However, the big benefit would come to big firms who could get an instant snapshot of their firm's position. Normally, their traders scribble their trades on order slips, which a clerk enters into the firm's computer system. In a fast-moving market, the information the clerk enters from the floor is obsolete. The new electronic hand-held communicators would enable the trader to transmit the information to the firm's system instantaneously, providing a real-time position for the firm. The ability to have this information provides decreased costs to the firm, and the customer gets better service and, hopefully, increased gains.

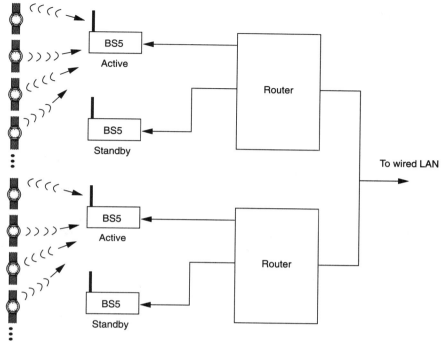

Figure 7.1 Typical Granite Market Maker network.

Wireless issues in financial services

You must be certain to provide proper security, such as encryption, when sending information over trading-floor wireless networks. Stock information is highly sensitive. No doubt you have heard about individuals like Milliken prosecuted for insider trading. You might have heard about the time a Bill Gates' internal memo criticizing Microsoft's performance cost him personally about $200 million. The tone of the memo caused Microsoft's stock to tumble at a great hardship to Bill himself.

If you had information about who was buying or selling what, you could leverage this information and make money. Stock exchanges take this seriously. One exchange, which runs fiber-optic cable between floors, also wraps the cable in a gas-pressurized, tamper-resistant conduit. Any attempts to get to the cable (which by the way, is extremely difficult to tap) will set off an alarm as the gas escapes. These people are serious!

So, it stands to reason, any wireless network carrying financial data will call for encryption at a minimum. There also will be a requirement to harden the antenna access points. You might require message authentication. But we're jumping ahead. We'll look at security in chapter 9.

Looking at the financial services industry, we might expect to see wireless used in the following ways:

- To link tellers' terminals within a bank's branch to a LAN.
- To provide print service for contracts or other official documents in a branch or office.
- To link self-service kiosks to the branch or office's LAN.
- To provide LAN access to traveling sales and support staff, or auditors.

Banks are prime candidates for wireless LANs. As they stress marketing and customer service, they are looking at ways to change the layout of their branches to accommodate their new focus. Wireless can help, even in old buildings (such as those found in Europe) that are heavily decorated with marble. Engineers can easily install wireless networks in these buildings in a matter of hours. They can even prebuild the network in a central location and test it before delivering the service to a branch.

Doubtless, other applications can be thought up. Maybe your organization has some unique wireless applications right now and is looking for more. When looking for good applications, look beyond wire replacement to breakthrough uses. You'll learn more about breakthrough applications in chapter 11.

Services Industry

This section of the book covers perhaps one of the most dynamic growth areas for wireless LAN technology—the service industry. Until the last two decades, providing service workers with equipment was not a high priority. Sure, you would see the occasional store or restaurant with electronic cash registers, but that was about the extent of automation. The amount spent on service workers was dwarfed by the amount spent on office workers. Possibly, retailers and other services thought labor was less expensive than equipment. However, we now know that this is not the case. Unmistakably, there has been a drive to automate the service industry. Let's look at how one segment of the industry, food services, uses wireless technology.

Food services industry

Competition in the restaurant business is fierce. Customers are looking for good value, high quality, and good service. Restaurant owners are looking for a way to satisfy the customer, which differentiates them and generates repeat business. The customer's main contact with the restaurant is the order taker who is constantly moving around the restaurant. Why are some restaurant owners, searching for ways to improve service to their customers, equipping the order takers with hand-held terminals?

Many restaurants use paging technology to notify the servers that their order is ready. However, notification of a ready order is only half the equation. The service improvement resulting from immediate placement of a clearly printed food order with the kitchen and the immediate notification to the order taker that the order is ready might justify the cost of equipping the restaurant with a wireless network. Increased efficiency of steps might even reduce the number of order takers required, while speeding service and delivering hot food to the table more reliably. For busy restaurants, the speedier service might even increase the number of sittings per day. It might even lead to higher sales per sitting. How many times have you gone to eat before going to a play or movie, and rushed off without having the dessert or coffee you wanted?

Food services wireless techniques

The wireless techniques for food service are identical to those for financial services. The hand-held communicator prompts the server with a restaurant menu on screen. The server enters orders on the screen using a stylus or keypad. The communicators send the data to the base station over a wireless, spread-spectrum radio frequency LAN. It transmits and receives data at data rates up to 121 Kbps and provides an RS-422 link to the host computer. Five to 15 servers can transmit fairly large packets of data at once.

The base station then relays orders to the bar or kitchen and the point-of-sale terminal or electronic cash register. The kitchen can respond instantly about the unavailability of certain dishes, rather than several minutes later, as is normally the case when an item runs out. Also, the bar can respond immediately to a request for a certain cocktail or brand of liquor.

Wireless benefits in food services

There are many tangible and intangible benefits to the use of wireless in the food services industry. Restaurant owners discover that they and their customers are happier because the billing operation is more accurate and controlled. Finally, the advantages in inventory control from recording the orders on a central computer are realized. The result is that the owner has happier customers and higher margins.

Wireless issues in the food service industry

In food services, there is a requirement for support of from 5 to 15 portable devices whose aggregate capacity demands are high and that provide full coverage throughout the restaurant. The issues for food services are not as severe as those for the health care or financial sectors.

Before leaving the services sector, let's briefly look at two other industry segments, namely:

- Hotel.
- Airline.

Hotel personnel can use wireless technology to check in and check out guests curbside, and enter and confirm guest requests for services throughout the hotel. The next time you are in the hotel, just use your imagination to figure out where potential applications will come. Maybe someone could figure out a way to update the hotel's accounting system using wireless when you help yourself to the minibar! That way, you can get an accurate bill when you leave in the morning, without waiting for the minibar to be verified.

Using your imagination in the airline industry uncovers some unique applications for wireless. For instance, maybe the long lines at the airport could be eliminated. Using wireless, you could access the reservation system and make your seat selection. This would dramatically reduce queues. Similarly, you could use a portable with a scanner to check in your baggage.

Today, airlines cannot expand their counter service without committing to costly "storefront" and without wiring the space. With wireless technology, customer representatives can take space as required, or even share space with other airlines on an hourly basis.

On the tarmac, everybody and everything could be connected to a wireless LAN. Cargo routing could be changed while cargo is being off-loaded. Maintenance workers could directly contact the required detailed information for servicing the aircraft. Using a scanner and wireless, in-flight meals could be confirmed or changed. In the air, airlines could offer passengers laptop computers connected to onboard networks. With one access point, the whole plane becomes "wired" without worrying about the weight of the cables. Let your mind go, and you can figure out potential network applications in the air: electronic mail gateways, office productivity tools, and video games, to list a few. There are a multitude of applications in and around an airplane that call for wireless LANs.

Retail Industry

For applications such as inventory control systems, the need for local area wireless data communication in retailing applications is synonymous with the growth of electronic point-of-sale (EPOS) devices connected to the central computer .

Retail wireless techniques

There are two types of EPOS devices: those that act as "online" terminals to the central computer, and those that can operate stand-alone in an "offline" mode. You can fit both with bar code readers. The online devices transmit each scanned bar code to a central computer and ask for the corresponding product description and price, whereas the offline devices have their own copy of this information. The offline devices only need to communicate with a central computer to record daily transactions and to update product/price information.

The system permits frequent sales analyses to ensure that stock levels are maintained and that nonmoving items are repositioned or scrapped. The method also increases customer service by speeding up the lines at the cash registers (especially when combined with online verification of debit and credit cards).

To further enhance customer service, some retailers have fitted shopping carts with a communicator incorporating a bar code scanner. Using this device, the customer can find out the location of a particular item he or she might want to buy, check the price of an item without having to search the racks or shelves for the correct price, and keep a running total of purchases. This device, like the electronic point-of sale devices, can be online or offline. The former would necessitate wireless communications.

Use of hand-held bar code scanners and wireless communications is likely during a retailer's regular stock taking. Figure 7.2 provides an example of a barcode scanner attached to a wireless transceiver, which could just have easily been a wireless portable or a hand-held communicator.

In another wireless application, sales staff use hand-held communicators to capture sales data. The communicator asks the salesperson key questions about product need and delivery requirements. The salesperson selects answers and choices from questions presented on screen.

The communicator is user-friendly, and personnel usually can be up and using the touch-screen, spread-spectrum device with a day's training. The communicator eliminates the need to master and memorize complex keyboard combinations or to cope with the slowness of an electronic stylus. One example from Granite Communications, the VIDEOPAD, features icons, graphics, and alphanumerics. The screen instantly places videobytes of information or commands under the control of warehouse personnel.

Wireless networks provide connectivity between the communicators and the LAN using spread-spectrum radio communications or infrared. The use of radio waves is currently the most developed and popular method for pro-

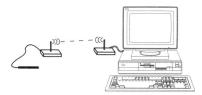

Figure 7.2 Wireless link to point-of-sale or stock-taking device.

viding wireless links in this niche. Radio-based LANs use radio wave transceivers in broadcast mode to send data from a communicator to the LAN. Most communicators also incorporate a barcode scanner, allowing error-free data capture and transmission.

Wireless benefits in retailing

Although wireless LANs hold little advantage over the evolving dedicated wireless EPOS systems, they can be used to provide flexibility in existing cabled systems. The frequent change in shop interiors calls for a flexible network. Many hand-held communicators can send and receive messages to the base station without interference or errors, and real-time connectivity to the database is assured.

Wireless issues in the retail industry

The first requirement for the LAN used to interconnect the real-time wireless devices is that it has a high availability, even in the face of high background noise from antitheft devices, neighboring wireless LANs (especially in shopping malls), fluorescent lighting, and other electrical devices. If electronic funds transfer is used as well, security of the network also is crucial. Demands for capacity and range are low, although the network might have to cover several stories in a building.

Examples abound of wireless applications in retail and merchandising. You also can find examples of the use of wireless in marketing. At any computer show like COMDEX or INTEROP, you will find wireless networks because large vendors like Apple and Microsoft want to showcase their network products. These corporations participate in hundreds of trade shows annually. A booth at a show might range in size from 20-x-30 feet to 100-x-100 feet, an ideal size for a wireless network.

At the INTEROP 93 August show, Windata's FreePort system provided wireless connectivity for 50 workstations in Microsoft's 50-x-50 booth. Using wireless helped Microsoft expedite the installation of the booth and reduce the cost of laying cabling. Because most shows are either set up or broken down in off hours (for instance, Sunday) cable installation costs can be astronomical. Union electricians must perform this service because most halls, salons, auditoriums, and ballrooms are union shops, and they charge overtime rates for Sundays. Moreover, a wired system placed restrictions on the location of workstations after the electricians finished. Wireless also eliminated previous setup nightmares from damaged and uprooted cables caused by forklift drivers. Because of these benefits, many vendors have found wireless to be an ideal solution for their temporary networks.

Speaking of temporary memories brings to mind Desert Storm. Armed forces can use wireless to establish temporary networks in the middle of a war. Two different divisions up to 2 kilometers away can share a network, as can personal computers within an encampment. Wireless is perfect for any temporary network: a new building, disaster recovery, marketing, and any other LAN on the run.

Wireless techniques in warehousing

A retailing application that's similar to stock-taking is warehousing. Warehouses generally use computing devices for stock control. Two types of devices are involved: devices to collect the data and computers to process it. Data acquisition is done mostly using hand-held devices, whose growth has paralleled the growth in the use of automatic identification technology (for example, bar coding).

Traditional devices store collected data, hence delaying the processing of the data until the device is brought to an access point where the stored data is uploaded to the central computer managing the inventory. Warehousing operations are moving towards stock management systems that can satisfy the demand for just-in-time delivery and real-time inventory control. Such systems need a constant, accurate reporting of inventory and data acquisition devices that can immediately transmit collected data to the central computer. This creates a need for wireless communications.

Wireless benefits in warehousing

Warehousing operations have a strong need for networks that can accommodate various mobile devices and fixed automated handling equipment. Flexibility of the network will facilitate changes to the warehousing layout, as might be called for when different goods are stocked.

The use of on-screen electronic signature capture can help confirm customer pickups or trace inventory to responsible personnel in real time. As a result, invoicing can be expedited and accountability strengthened, providing happier customers and greater control.

Wireless issues in warehousing

The capacity demands in warehousing are low, but the coverage area can be wide, although generally less than for industrial applications.

Dedicated wireless portable data acquisition systems have been commercially available for 15 years. Organizations could nevertheless benefit from wireless LAN technology where it can provide interoperability of portable, mobile, and fixed devices on a common medium with increased support for more devices.

Now, you have an appreciation for local area networking, wireless LAN technology, and innovative applications from several industries. As mentioned, the various WLAN technologies have their various niches. Table 7.3

TABLE 7.3 Summary of Wireless LAN Approaches

Type	Speed	Usage	Pros	Cons
Infrared	2 Mbps	Line-of-sight data transmission, peripheral-sharing devices	Ease of installation and use, mature technology; permits faster transmission than many other wireless LANs	Line-of-sight transmission only, units aligned and data path kept clear
Microwave	5.7 Mbps	Short-range data transmission	Ease of installation, good compatibility	Signal cannot penetrate all walls, inappropriate for highly mobile sites
Spread-spectrum radio	2 Mbps	Short-range data transmission	Signal capable of penetrating supporting walls, flexibility	Unreliable near heavy electromagnetic transmissions, requires one transmitter per PC

shows a summary of the pros and cons of wireless. So, when you start the initial design of your network, refer to this chart to refresh your memory about infrared, microwave, and spread-spectrum. In addition to the pros and cons in Table 7.3, there are some other considerations such as health and safety, security, and network management. In the next part of the book, you'll pick up information about these key issues.

The Problems

"The fantastic advances in the field of electronic communication constitute a greater danger to the privacy of the individual."

Earl Warren

"We must make the best of those ills which cannot be avoided."

Alexander Hamilton

"The economic and technological triumphs of the past few years have not solved as many problems as we thought they would, and, in fact, have brought us new problems we did not foresee."

Henry Ford II

8

Health and Safety

Usually, people object to wireless data communications technology for two reasons. The first reason is security. These people object because they believe that anyone can capture wireless transmissions and decode messages. But this is really not a problem introduced by wireless networking, as you'll see in the next chapter. The second reason cited by many detractors of wireless is the health and safety issue. Some people find the thought of microwaves zipping through the air, and their bodies, a little disconcerting. They believe that the radiation from the microwaves will harm them. I mean, would you put your head in your microwave oven? In truth, it is the electromagnetic fields (EMFs), not the radiation, that should worry you.

Our Electrical Environment

Electromagnetic fields are found everywhere that electricity is in use: around power lines, office machinery, wall wiring, lights, and video-display units (VDUs). Electromagnetic fields have been around in great abundance since 1882, when Thomas Alva Edison set up the first electric power transmission plant in the United States. It has been more than 110 years since electricity generation started, but only 65 years since radio transmissions began, and 45 years since radar and telecommunications entered our environment. Like natural fields, human-made fields are limited by the physical properties of the environment. Unlike natural fields, they can interfere with our bodies and cause harm.

For a long time, nobody paid attention to these omnipresent, invisible fields because they did not think them harmful. Then, in the early 70s, research began to establish some worrisome connections between EMFs and serious health problems, including cancer, and especially leukemia. The past five years have seen a revolution in public awareness of electromagnetic fields, including computer equipment, and especially fields produced by cathode-ray tubes (CRTs), the monitors used with nearly all desktop computers.

Researchers have not established an ironclad link between exposure to electromagnetic fields (from radio transmitters, CRTs, overhead power lines, or other sources) and ailments such as leukemia, but there is enough circumstantial evidence to worry many people.

Imagine a World without Electricity

So, what are EMFs? To understand that, you need to understand a little about electricity. You live in an electrical environment. As you go about your business, just about everything you do gets an assist from electricity. At every point in your daily routine, you see evidence of a dependence on electricity. You probably awoke this morning to the alarm from your clock radio. After rousing yourself from sleep, you trundled off to the bathroom, where you turned on the light and plugged in either an electric razor, hair dryer, electric plaque remover, or electric toothbrush. Heading for the kitchen, you turned on the light, popped some bread in the toaster and turned on the coffee maker. You get the picture. Electricity is all around you, in more ways than you know. But it doesn't stop in the home. When you got to the office, you used a photocopier, a personal computer with a CRT, and a fax or a phone with call display. And that's not all. Don't forget the fluorescent lighting, the microwave, and the soda machines in the employee lounge. You returned home on the electric-powered subway and turned on the oven, the microwave, the television with the remote control, or the stereo with the infrared headphones. From the time you wake up in the morning to the time you retire at night, you are in constant contact with some sort of electrical device.

The Invisible World

You cannot (nor can anyone else), see electricity so you can only visualize it by looking at the effects it produces. As you are aware, everything is composed of atoms. An electric charge is a basic property of certain particles within an atom that causes them to attract or repel each other. There are two types of electric charges: negative and positive. Electrons carry the negative charge, while protons carry the positive charge. As you know from love and magnets, opposites attract, and likes repel.

This attraction actually binds electrons in an orbit around the nucleus of each atom, and, on a larger scale, binds atoms and molecules together. An ordinary atom is electrically neutral because it contains the same number of electrons and protons. You can upset this balance by adding or subtracting electrons to create a charge. Electrically charged objects exert a force on other objects. You can harness this charge and potential force as electricity. Experts measure this potential force or electric energy as a volt.

Measuring Electricity—Volts, Amps, and Watts

The effect of different charges on each other produces the voltage, or potential electrical energy, in an electrical circuit. Current is the actual flow of the electrical charge through the line. Most people think of voltage as the power in a line. But this is not exactly accurate. Scientists and engineers measure electrical energy in volts (V), current in amperes (amps), and electrical power of a line in watts (W). You can obtain the wattage of a line by multiplying the voltage by the amperage.

In North America, we use a 60-hertz (Hz) alternating current (ac) power system. Alternating current simply means that the electrical current running through the power lines and wall wiring alternates back and forth on itself. Hertz is a measure of alternating current, indicating how many times per second the current cycles, or changes direction. For example, 60-Hz current reverses itself 60 times a second. Many of our spread-spectrum bridges reverse themselves between 902 and 928 million times a second.

Electricity travels through the air (lightning) or along wires or conductors that allow electric charges to flow through them. Living things are good conductors. Any time an electric current runs through a wire, the air, or an appliance, it produces an electromagnetic field. You find electromagnetic fields wherever there's electricity, and around any object that has an electric charge. All electrical devices produce electromagnetic fields. Electromagnetic radiation is the wavelike fluctuations of the electric and magnetic fields produced when the electric charge accelerates.

EMF Emitters

In today's electrical environment, EMFs are everywhere: around power lines, radio and transmission towers, electrical outlets, office machines, and CRTs. Let's look at the potential EMF emitters at our wireless workstation. First, you have the radio transmitter, computer, and the CRT. CRT fields can be more intense because of the high voltages used to fire electrons from the rear of the monitor on to the screen. The yoke, a network of electrical cables that surrounds the picture tube, ensures that the electron beam goes where it's supposed to and generates a strong field. So does the power cable. In the case of our spread-spectrum bridge, the fields it produces are called 902-MHz electromagnetic fields.

Electromagnetic Fields = Electrical Fields + Magnetic Fields

Actually, electromagnetic fields consist of electric fields (EFs) and magnetic fields (MFs). Normally, they are produced together by electricity. They generate radiation in the form of waves, and they drop off as you move away from the source.

Electromagnetic fields and dosages are abstract units like teslas and greys and are difficult to visualize, making an already complex issue even harder to grasp. However, one easy-to-understand scientific fact is often overlooked: field strength varies with the inverse square of the distance between the source and the measuring device.

If you are standing 1 meter (about 1 yard) away from a source and then move to 2 meters away, you reduce the field strength to ¼ of its original value. Move 8 meters away and you reduce the field to ¹⁄₆₄ of its original value, and so on. Thus the easiest way to reduce your field exposure is to move away from the transceiver or network device. But we digress. So back to EFs and MFs.

Electric fields are a property of electric charge at rest, magnetic fields are a property of electric charge in motion, and radiation is a property of electric charge being accelerated or decelerated. Electric and magnetic fields are actually quite different, and they have different effects on living things. When EMFs interact with humans, the electric and magnetic fields uncouple and affect living organisms separately. Electric fields are shielded by many things (for example, trees), whereas magnetic fields pass through anything that isn't mostly iron. It is this uncoupling that might cause the real dangers of EMFs—exposure to magnetic fields.

The Electromagnetic Spectrum

To get a better understanding of electromagnetic fields, you need to understand electromagnetic energy or radiation. In Figure 8.1, you can see the electromagnetic spectrum covering the wide range of electromagnetic radiation or energy. Scientists arrange the spectrum according to frequency and wavelength of the various types of radiation. As you learned in chapter 3, frequency is the waves per second sent out by the energy source. Wavelength is the distance that a wave can travel before it flows back to its source, say, our radio transmitter. Higher-frequency radiation has more waves of shorter wavelength, and vice versa.

At the high end of the spectrum come the gamma rays, X-rays, and ultraviolet rays, called ionizing radiation. These rays have enough energy to enter cells and break chemical bonds, to kill instantly, if exposure is high enough, or to cause cancer. While your CRT emits ionizing rays like X-rays, CRTs are shielded well enough that such emissions are not a major issue. Health hazards, if any, are attributable to electromagnetic emissions, not X-ray radiation.

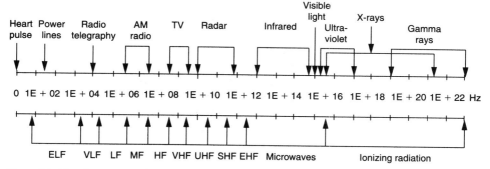

Figure 8.1 The electromagnetic spectrum.

All other forms of radiation are nonionizing, which means they don't have enough energy to get inside your cells and break down the chemical bonds. This doesn't mean, however, that nonionizing radiation doesn't have bioeffects.

Nonionizing radiation is broken down into middle and low frequencies. Middle frequencies (known collectively as microwaves) come from visible light, heat, radio frequency, radar, and television waves. Scientists and bio-medical specialists know that microwave radiation is dangerous because of its thermal effects (it can heat or cook tissue). It also has nonthermal effects (it can cause certain biological changes). Broadcast transmissions, radar, satellites, CB radios, RF transmitters, electrical security systems, telephone switches, sonar, CRTs or VDUs all emit microwave radiation. Appendix B shows where these devices fit in the electromagnetic spectrum. At the lower end of the spectrum is extremely low frequency (ELF) radiation around 60-Hz power lines and appliances. Recently, scientists have discovered that ELF fields can cause cancer and other diseases.

Your Body Is an Antenna

So what does all this mean, and what do these electromagnetic fields do to you? It means you absorb a lot more energy than you realize, and this energy has some dangerous biological effects on you. As previously mentioned, you absorb energy because you are a wonderful conductor. Scientists divide all matter, including humans, into conductors, which can transmit electricity by allowing it to flow through, and insulators, which do not transmit electricity. Stand in an electromagnetic field, and you act just like an antenna.

In addition, your body has a higher conductivity than the air. This means that an electric field only will produce some small surface currents on your skin, and a limited amount of radiation will enter your body. How much depends on your size, body shape, grounding, and the field. So, each time you touch an electric appliance, such as a photocopier, little contact currents run from the photocopier to your body. Except for potential shocks, there is little concern about these electric fields.

On the other hand, exposure to magnetic fields is another matter. The human body has a magnetic permeability almost equal to air, so the whole magnetic field will enter it. The difficulty is you probably won't detect exposure. Right now, you might be sitting amidst a strong magnetic field and not realize it. Today's high-technology, state-of-the-art, electronic workplace is a sea of electromagnetic fields.

Be Proactive, Not Reactive

But even if you aren't convinced that there is a link between electromagnetic fields and health and safety, you might want to follow the example of Blaise Pascal's wager on the existence of God. Pascal argued that because you cannot prove there is or isn't a God, you should err on the side of caution and act righteously, as if there were a God. Because it's relatively easy to reduce your exposure to electromagnetic fields, you are unlikely to be worse off doing so, and you might be much better off. Because it is impossible to avoid these fields, distance yourself from them. If you are using infrared or microwave technology, then you should consider the placement and siting of these devices. Also, remember to turn off your machines when you are not using them. Magnetic fields go away when the machine is off. Most people run their workstations, servers, hubs, and bridges all the time, even when they are not used. If your organization is concerned about EMFs, then purchase a gaussmeter, measure the office areas, and take precautions.

Finally, the best defense against EMF is education. This chapter provides you with the basics of electricity, the electromagnetic spectrum, and electromagnetic fields. When you are purchasing wireless equipment, ask your vendors what they have done to protect you and what they recommend for safe operation of the equipment. Ask questions and protect yourself. The theme of this chapter has been that an ounce of prevention is worth more than a pound of cure. In the next chapter, we'll continue this theme, and you'll see another problem to protect yourself from—unauthorized disclosure of information being transmitted.

9

The Security of Wireless LANs

Now that you've seen some actual examples of wireless LANs in action, you might be thinking about how they can be used in your organization. But there's probably a question in the back of your mind, a major concern that might even be holding you back from using the technology. It's something you need to know before you are willing to commit your company's business information to a wireless LAN. Are wireless LANs secure?

If you haven't wondered about security, you should. It's an important consideration for any component of an information system, be it hardware, software, networks, or the data handled by each. That's because each component is an asset to your company, and as a manager or staff member, you have a responsibility to protect those assets. You must take the steps necessary to safeguard your computer systems and information. But the steps must be cost-effective. You shouldn't spend more on securing assets than the likely cost of recovering the assets in the event they are lost. Security is a balancing act between the value of information assets and the cost of protecting against the possible threats to them.

From what you've learned so far, your intuition probably tells you that wireless LANs pose some unique security concerns. On the other hand, you might have heard the claims of some that, "Security in wireless LANs is not an issue."

Well, your intuition is right. Anytime you use a medium that involves sending messages so that everyone else can hear them, you can bet that security is an issue. Even a wired LAN faces this problem because every node on a given LAN segment can capture any data placed on the cable (see Figure 9.1). However, the use of electromagnetic radiation as the medium

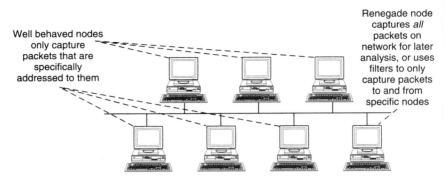

Figure 9.1 Capturing data from a LAN.

for wireless transmissions increases the risk because the signals are not confined to a wire. Instead, they spread out through space, limited only by their own transmission characteristics such as the ability to penetrate solid objects.

(While it's true that cables also can allow signals to escape, that's not a necessary condition for them to function, as it is for wireless. Various techniques like shielding can also be used to limit the extent of radiation. Further, media such as fiber-optic cables have practically no extraneous radiation.)

When security is declared a nonissue, it's usually based on some of the old, reliable fall-back excuses that have been used to deflect security concerns for many years and for many different kinds of information technologies. Examples (and their translations) include the following:

- Security is "built-in." The system's specifications are secret, so nobody knows how it works. (We really believe no one else is as smart as us, and that "security through obscurity" is the best technique).

- It's just too challenging, or difficult, or expensive to acquire and use the technology necessary to break into the system (at least as far as we know, and at this point in time, what people will come up with tomorrow is anyone's guess).

- All the hardware is designed so that it only operates in an authorized and secure manner (as long as you and everyone else only uses our hardware and promises not to modify it).

- It would be illegal for anyone to try to break our security. (We believe that laws are totally effective deterrents. We also believe that all this talk about system cracking is just media hype, and that there is a tooth-fairy).

- The system isn't used for anything that important, or valuable, or secret (so please don't ask us why we bought it).

Most of these justifications simply are excuses for not taking positive steps to secure the company's systems. It usually is much easier to try to explain away security as a nonissue.

However, if these rationalizations are all true, then no one has ever listened to a mobile telephone call, or watched a scrambled television program from a satellite broadcast or cable television system, or used a cellular telephone to place a call that is billed to another party, or captured a logon password as it passes by on a LAN cable, or even discovered a juicy tidbit of information on a supposedly boring office electronic mail system.

Of course, each of these security breaches has happened, probably lots of times. The proof is available through documentation or even demonstration. (Nothing makes a believer out of someone better than actually grabbing a password or letting them hear a cellular phone call). After the breach occurs, the same people who came up with the original excuses might try to convince themselves that the problem is very limited or confined. Again, its easier to explain away a problem than to do something about it.

As you see in this chapter, wireless LANs do face security exposures. In some instances, their advanced technology might make the task of breaking security more difficult. But the exposures are still there, and they still need to be addressed. To claim otherwise is simply irresponsible.

Fortunately, the exposures can be addressed. The best way to deal with them is to understand them and to deal with them in a prudent manner, using some of the techniques you'll see later. After learning more about the risks, there might be some highly secure applications that you won't want to put on a wireless LAN, or any shared medium for that matter. However, in most cases, you'll find that just about any problem can be addressed. Taking this approach allows you to enjoy the benefits of wireless technology without accepting an undue risk to your business.

The Need for Security

First of all, what do we mean by security? There are many definitions for the term, and almost all are valid. The variations simply represent different perspectives, and in some cases, just different semantics. Most have, at their core, the following common concepts:

- Information assets (including hardware, software, and data) that require protection.
- Security goals or security objectives for those assets.
- Threats to the security goals for the assets, also expressed as risks, where the goals are not achieved (sometimes quantified in terms of the probability of a threat or risk causing an expected dollar loss).
- Security measures (also known as counter measures) that help address the threats or risks.

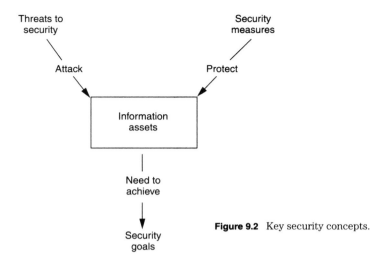

Figure 9.2 Key security concepts.

Figure 9.2 illustrates the relationship between these concepts.

Information assets

One important duty for a company's management and staff is to properly safeguard the company's assets. This doesn't just mean locking up the cash and inventory. For computerized information systems, it extends to information assets, which include:

- Hardware (such as, computers, printers, disk and tape drives, and communication equipment, including network components such as routers, bridges, and hubs).
- Software (such as operating systems, database management systems, production applications, and network programs on servers and clients).
- Data (such as application databases, configuration files, and information in transit across a network).

As you can see, the definition of information assets is quite broad. It includes both tangible and intangible assets, all requiring protection.

Security goals

What do we mean by protection, and what is it that we are guarding against? A commonly held view is that information must be protected against the loss of three key characteristics: confidentiality, integrity, and availability. Maintaining these characteristics is the goal of information security.

Confidentiality. You have information confidentiality when the data involved can only be viewed by an authorized person or process. Sensitive data, such as customer files or payroll details, needs to be kept out of the wrong hands. If confidentiality isn't maintained, the company might lose competitive advantage, or be subject to statutory or legal remedies for not protecting privacy.

Protecting confidentiality normally involves preventing access to the data, for example, by using operating system mechanisms to revoke read rights to a file from an individual. Another approach is to encrypt the data using a high-quality cryptographic algorithm.

Integrity. Loosely speaking, if information has integrity, it is correct. More precisely, the data is complete, and accurate, as well as being fully up-to-date (known as being timely). Depending on your outlook, a quality such as "accuracy" might mean that the data is absolutely, 100 percent exact, or it might be correct within some defined limits. Security plays a key role in protecting data integrity, by guarding against unauthorized changes to information.

A technique such as using file access controls can help ensure that only authorized people or processes can make changes to information (that is, adding records, modifying existing data, or deleting files). Cryptography also can help ensure data integrity, through techniques like cryptographic checksums and digital signatures (more on these later).

Note that ensuring the integrity of information requires more than just proper security. Errors can creep into a data file, even if changes are made only by authorized staff. This is where traditional accounting controls come in, like transaction balancing and edit checks, which are designed to prevent and detect errors. The debate as to whether information security is part of good accounting controls, or vice versa, is left to you and your auditors.

Availability. For information to be of any value, it must be available when it is needed. Otherwise, all the money spent on building systems, and on collecting, processing, and reporting data is wasted. Loss of information availability can occur because of hardware failures, program bugs, data file corruption or destruction, and many other reasons. What they have in common is that they can leave information unavailable when a key decision must be made.

Security controls can help prevent a loss of availability, through (once again) proper file access controls. For example, by ensuring that only a few people can delete a file, there is less chance of accidentally erasing the data. Other techniques, such as regularly backing up key data files, and developing a Business Recovery Plan (BRP), help recover from problems that can make systems and information unavailable.

Threats or risks

There are hundreds, or thousands, of different kinds of threats to information assets (though threats can be placed in a few threat categories).

Everything from a fire in the data center, to a cracker breaking into the network, to an employee accidentally deleting a file, all represent instances where a security goal has been violated for an information asset.

If a threat actually occurs that violates a goal for an asset, then there usually is some idea of loss. The loss might relate to the direct costs of recovering the asset. For example, when a mainframe computer is destroyed by fire, or if a data file is lost, it must be reconstructed from scratch. Or the loss might be less tangible but equally costly (for example, when a confidential customer file falls into the hands of a competitor, or a file of outstanding sales invoices must be reconstructed from original documents).

Security measures

Security measures are the steps, tools, and techniques you use to help prevent threats from actually happening, or to mitigate losses if they do occur. Examples include:

- Physical security barriers protecting a computer room.

- Operating-system-based mechanisms such as user-IDs and passwords, and file access controls.

- Communication and network facilities like call-back modems, firewalls, and link-level encryption devices.

Rarely can you rely on a single all-purpose security measure to provide the needed protection. A well-constructed security system will use many different types of measures in a layered fashion, most trying to head off trouble, while others limit the damage that occurs. Taken together, they should provide the most cost-effective approach to protecting the company's information assets. See Figure 9.3.

Security measures also carry a cost. Administering user-IDs and file access permissions for a computer operating system takes up the company's resources, just as does purchasing locks or hiring a security guard. The trick is to balance the cost of security measures against the cost incurred when security goals have not been met for information assets.

Are Wireless LANs Secure?

Now that you understand the need for security, and the key concepts involved, we can return to the main question at hand: are wireless LANs secure? If you hate long, convoluted explanations filled with the typical hedging, back-peddling, assumptions, limitations, clarification, and elucidation, then the simple answer is "no." But then the simple answer rarely does the subject justice, and so it is with wireless LANs. Let the hedging begin.

First of all, the question implies an absolute level of security. That is, is

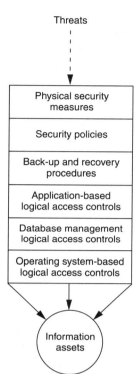

Threats

Physical security
measures

Security policies

Back-up and recovery
procedures

Application-based
logical access controls

Database management
logical access controls

Operating system-based
logical access controls

Information
assets

Figure 9.3 A layered approach to security.

there any chance at all that security can be breached by any possible cause at any time now or in the future? Absolute security likely is impossible to achieve. There is no current system that can guarantee with 100 percent certainty that all security risks have been eliminated, or nullified so that they will not occur. Even if such a system were possible, it would not be desirable. For most business systems, absolute protection is more than what is required. This means that more security is purchased than is necessary, which is not cost-effective. So security must be judged against the threats that are most likely to occur, or that can cause the most damage even if they aren't as likely.

Second, the question of whether something is secure cannot be answered unless all the security measures that are in place and working are known. All computerized systems face threats. Without the use of adequate counter measures, there is no security. Consider the types of security measures that are available, but evaluate the ones that are actually used.

Third, when the question is asked in the context of wireless LANs, it almost certainly is meant relative to wired LANs. Most everyone understands that an information system running on any LAN faces security exposures.

For example, hosts and file servers connected to the LAN call for user-IDs, passwords, and file access controls. If these measures aren't in place, security goals can be violated. This is equally true, regardless of whether node connections are wired or wireless. What we really want to know is whether wireless LANs face any additional security threats, compared to wired LANs.

So after some deft footwork, you find you must rephrase the question: "Do wireless LANs pose any additional threats to our information assets, relative to wired LANs, for which no security measures exist, and which might result in a violation of our security goals?" Wordy, but much more precise, and something we can deal with in the following sections. You'll see the unique security exposures faced with wireless LANs, and some of the techniques available to effectively counteract them.

Wireless LAN Security Exposures

As we mentioned earlier, wireless LANs face all the same security exposures that wired LANs do. You still need to be concerned about security measures like:

- Controlling access from the network to important information and services offered by host computers, file servers, and database servers. Most popular host and network operating systems or database servers offer a full suite of security mechanisms that identify and authenticate users, control what data they can access, and monitor attempts at unauthorized activity. When a security breach occurs, the most likely problem is that these mechanisms simply haven't been activated or configured correctly.

- Ensuring that resources on workstation nodes are protected from unauthorized access and use. Unless additional security steps are taken, most workstations will give full access to hundreds of megabytes of stored programs and data, simply by being powered on. In some cases, they might even automatically login to a network server. At the very least, workstations should have an initial password for access. Consideration should also be given to additional security programs.

- Providing proper physical security over network nodes. This doesn't just apply to high-profile network nodes like file servers and workstations. It also includes network devices like routers and bridges. Besides having value as a physical asset, these nodes are often focal points for the entire network infrastructure. As such, they are ideal targets to gain access to network services or network traffic.

Each of these areas is crucial to LAN security, whether or not wireless technology is involved. However, using wireless LANs can increase the security exposure in certain areas, as you'll see in the following sections.

Theft of equipment

Physical security for computers, particularly today's breed of personal computers, always is a problem. Steps have to be taken to keep an inventory of equipment as well as to establish responsibility for each item. Building security procedures also should be in place to lessen the risk of computers being taken out without authorization.

Wired LAN nodes certainly can be stolen (the network interface card, the computer, the connectors, everything). So theft isn't a concern that's unique to wireless LANs. However, a major reason for going wireless is to give users mobility, typically involving notebook and subnotebook computers. Thus, you already have an environment where equipment moves around a great deal, and things can go missing very easily. There is a much greater risk of losing a wireless NIC, as well as the computer that surrounds it. Your inventory procedures must rise to the challenge or be prepared to process many insurance claims.

Eavesdropping/interception of messages

When a medium is shared, someone might be listening. Have you ever used a "party line," a telephone line shared by more than one subscriber? Convention and courtesy dictates that if you pick up the telephone when the party line is in use, you immediately hang up. But did you always hang up right away? You also only are supposed to pick up the phone when you hear your own unique ringing sequence, like two shorts and a long. Did you ever listen after two longs? Did you ever think your neighbor might be listening to you, too?

Eavesdropping or intercepting messages is not unique to a shared telephone system. A wired LAN is also susceptible because every node can detect every message sent on its portion of the network. Once this occurs, it is easy to lose the confidentiality of key data being passed over the network, whether it's a password entered during logon to a file server, or a payroll file update.

While NICs are supposed to behave themselves and only pass on the traffic intended for their computer, they can be made to collect all the data passing by or through them, sort of like getting your friend to listen on the party line and tell you what's going on. Or, instead of using a run-of-the-mill NIC, you can attach special network monitoring equipment designed for data analysis and problem diagnosis. In either case, you will be able to collect every packet on the LAN segment.

The difference with a wireless LAN is that, to intercept traffic, you don't have to physically attach to anything. Depending on the type of wireless transmission used, the signal can be available from a few yards to a few miles away (or from a few meters to a few kilometers in System International countries). All you need is the capability to receive the signal, and you're a good way there.

Even though the signal might be available over a greater area, some advocates of wireless technology claim that eavesdropping is more difficult on a wireless LAN. The theory is based on the fact that wireless LANs use transmission techniques that are more difficult to interpret, particularly for someone using generic receiving equipment. Without knowledge of the specific techniques used, it is difficult to translate what is intercepted into meaningful data.

It certainly is true that wireless transmission techniques make interception more difficult, especially when the LAN uses proprietary or unique ways to send data. The increased difficulty does help to diminish the security exposure. However, a motivated attacker will find a way to listen in. For example, depending on the specifications involved, an attacker could probably find out the needed information from the manufacturer. An attacker also doesn't have to start from scratch using general-purpose receivers or monitors. It probably is simpler to start with a wireless LAN manufacturer's standard NIC and modify it through hardware or software to deliver all the data it hears.

Overall, loss of confidentiality is definitely a greater concern on a wireless LAN than on a wired LAN. Later, you'll see some additional considerations and techniques that address this exposure.

Modification/substitution of data

Modification or substitution of data involves altering a message going between nodes, or suppressing the message and inserting substitute data. It's a tough trick on any LAN, precisely because of the use of a shared medium. It's like a room full of people who are conversing. One person can't change someone else's words as they speak, or steal their words and substitute their own. Similarly, a LAN node normally cannot modify the data from another node, or take if off the shared medium and insert its own instead.

However, recall our room full of people. Suppose Person A always waits for Person B, whom they are standing beside, to begin speaking. As soon as Person B starts to speak, Person A begins performing an extremely loud impersonation of Person B, so that they drown Person B out. Others in the room can only hear the louder voice of Person A, but they think that the words they hear come from Person B. In the same way, a LAN node using greater power could drown out another, thereby issuing false data. If an attacker modifies or substitutes data as it passes between nodes, then the integrity of the information is lost. See Figure 9.4.

Substitution of data in this way could be attempted on a wireless LAN more easily than on a wired LAN. Using additional power or a directional antenna easily could allow one node to overpower another. Wireless nodes are accustomed to dealing with different signal strengths, and they wouldn't be surprised to hear one node louder than others. The stronger node could

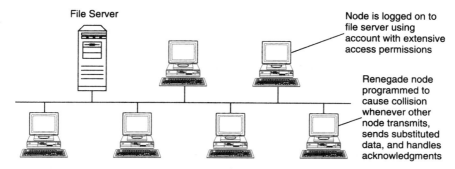

Figure 9.4 Substituting data on a LAN.

block the weaker one, substitute its own messages, and even acknowledge responses from other nodes.

Another area where modification or substitution can become a problem is at points in the network where traffic becomes concentrated. For example, you use a bridge to pass data between two segments of a LAN. Because every message between the two segments must pass through this one device, it represents an ideal place to modify traffic. The bridge could be used to modify a message as it's passed between the two LAN segments, as shown in Figure 9.5.

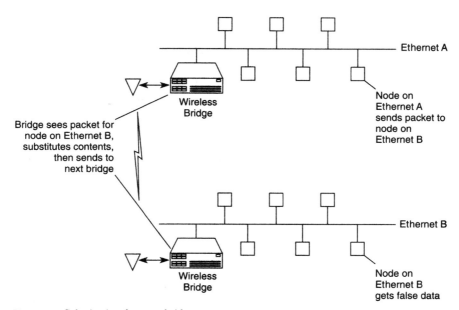

Figure 9.5 Substituting data at a bridge.

Wireless LANs are often connected to wired LANs through a bridge, or they depend on a central hub to pass messages between nodes. These devices make good targets to alter traffic passing between wireless nodes. While complicated, modification or substitution could affect the integrity of data on your wireless LAN.

Masquerade

Masquerade involves one node pretending to be another. While this also happens during the substitution of data, a masquerade takes place more easily because the node being impersonated is not around. For example, the computer might be shut off, or it might be still on but not connected to the network. Because the impersonated node isn't sending messages, the masquerading node doesn't need to worry about having to block other transmissions. By altering its own identification, it simply claims to be another node.

The motivation for a masquerade is that some network services are allowed or denied, based on the address of the node making the request. For example, the "rlogin" service available on many UNIX computers can allow users to login without a password, as long as they are coming from a specific network address. A more serious case would be when the masquerade took place after a user logged on to a powerful account on a host or file server. Say someone logged on to a Novell NetWare file server as the all-powerful SUPERVISOR. Disconnecting or turning off the user's workstation, or even disconnecting the user's segment of the LAN, then masquerading as the user's node, could permit unauthorized access to sensitive network commands or files. Both confidentiality and integrity can be seriously impaired.

Perpetrating a masquerade is arguably easier for a wireless LAN, once again because it is not necessary to physically connect to the medium. Thus, an attack could be mounted from any node that is in range of the wireless LAN operating area.

The argument still could be made, as it was for interception, that the increased complexity of wireless LAN transmissions makes masquerading as another node more difficult, especially if using generic transmission equipment. Protocols that tie network address information to hardware specific details (such as a unique number programmed into the NIC) make the task even more difficult. But while masquerading might not be trivial, it can be done. For example, a viable alternative is modifying a wireless LAN node purchased from the same manufacturer as the node being impersonated.

You might not have to worry about a wireless node trying to pass bad checks while posing as someone else. But masquerading as another node to access network functions or data is possible and needs to be countered.

Interference/jamming

Going to all the trouble of monitoring data from your wireless LAN, or actively trying to change it, or pretending it comes from another source, are intentional acts designed to breach security. However, it might be a purely unintentional act that gives you the most headaches—interference from other sources of electromagnetic radiation.

Noise or other forms of interference can prevent nodes from hearing each other, disrupting all transmissions and leaving information systems totally unavailable. Depending on the type of wireless LAN you are using, interference from many sources can affect you. Another wireless LAN in an adjacent office can hinder your LAN. Then again, so can the office microwave oven. The interference drives up error rates, resulting in slower network throughput as messages must be resent. At some point, communication between wireless nodes might stop altogether. See Figure 9.6.

An intentional act of interference, known as jamming, involves deliberately generating electromagnetic radiation to disrupt communications. The effect is the same—no LAN at all, or at least decreased performance.

Due to errors it can induce in wireless transmissions, interference or jamming can result in data integrity problems. However, you can address these threats through error-checking codes routinely used with wireless communications, since noise and interference always can crop up. With interference, the main problem to contend with is normally the lack of availability of the network.

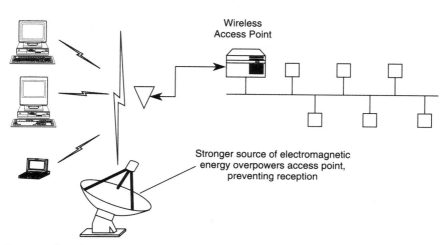

Figure 9.6 Interference from an outside source.

Security Measures

The exposures just described might paint a pretty dreary picture of the security of wireless LANs. But it's not intended to scare you away from using wireless technology, only to make you aware of the areas that need to be addressed. Fortunately, there are security measures that can address these exposures. Some cost more than others, or are more trouble to implement, but the right combination of tools and techniques can provide the level of security that you need in order to be comfortable with your wireless LAN. Let's look at what you can do to protect your information.

Shielding radiation

Earlier in the chapter, we conceded that the transmission techniques used for wireless LANs can make it more difficult for the casual attacker, one who is not prepared to go to a great deal of trouble. (We also recommended that you don't rely on this for security). However, you can exploit other transmission characteristics to your advantage to help in decreasing your exposure to threats like interception, modification, or masquerade.

As discussed in chapter 3, certain frequencies of electromagnetic radiation are not good at penetrating objects. Light is the best example, which will not pass through something solid like a wall. Microwaves have difficulty passing through steel or concrete. Therefore it is quite possible to effectively shield radiation so that it can't escape a confined, and hopefully secure, area. Shielding also restricts the signals that can enter the area. (Attackers have an advantage in that they could try to overcome the shield by saturating the area through increasing their power levels. However, if the shield is effective, they will not be able to hear your nodes in order to get the two-way dialogue necessary with LAN protocols). Figure 9.7 illustrates this.

You can use this characteristic to your advantage if you don't need a wide area of coverage for your wireless LAN. For example, when you only want to connect nodes within a meeting room or training facility, an infrared wireless LAN will function very effectively, and no radiation will escape from the room. Of course, this also helps prevent signals sent by active attackers from getting in.

Similarly, the radiation from a microwave LAN will have difficulty escaping from a heavy structure like a basement or factory. The type of antenna used (directional or omnidirectional) and its orientation also help limit the area over which signals radiate.

Even spread-spectrum LANs using relatively low-frequency radio waves can effectively be shielded by buildings that are below grade. (Try using a cellular phone from an underground parking garage.) However, this type of LAN is normally the most difficult to shield because it is designed for use throughout an area like an office complex.

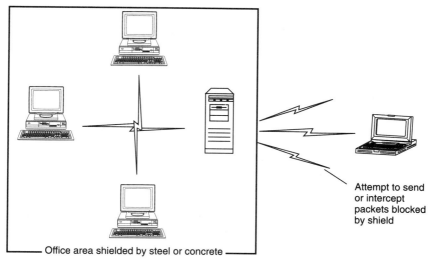

Figure 9.7 The benefits of shielding a wireless LAN.

The easiest way to determine how effectively a wireless LAN is shielded is to try using a node outside of the main operating area. If it can't participate in the LAN from outside, but works fine when closer, you are probably safe. For greater certainty, an electronics engineer or technician with experience in radio and microwave communications can use measuring devices to determine the field strength of your wireless LAN signals at various points outside of the operating area.

Security features in network operating systems

Since wired LANs also are susceptible to attacks like eavesdropping and masquerading, many network operating systems offer mechanisms to enhance the security of key data passing over the network. These features become even more important when using wireless LAN technology.

For example, passwords are particularly sensitive items of information because they provide the means for authenticating users. A user's password is an inviting target, especially if the user has a privileged account. Capturing a password as it goes across a network is fairly easy to accomplish with readily available tools.

In response, Novell's NetWare, like most network operating systems, encrypts passwords as they pass between the user's workstation and the file server where they are logging in. (You'll learn more about encryption a bit later in the chapter). In this way, even if the communication is monitored and the password is captured, it cannot be used to break in to the user's account on the file server.

Another useful feature offered by NetWare is packet signatures. This mechanism, based on a cryptographic technique, verifies that each packet received by a node has come from the node shown as the sender. Attempts to masquerade as another node are thereby defeated.

Even more involved mechanisms can greatly enhance the security of any LAN, particularly wireless LANs. For example, Remote Procedure Calls (RPC) is a protocol developed by Sun Microsystems, originally for use with their UNIX computers but now licensed for use with many types of machines. RPCs allow a process on one computer to request that another computer perform a particular function. For many of these functions, the identity of the computer and user making the request must be verified. Secure RPC is an extension to the protocol that allows each node to authenticate the identity of the other. It even supports encryption of all messages sent between the nodes over the network. When in use, no other node can intercept any meaningful data, substitute its own data, or claim to be one of the communicating nodes.

Encryption

We've mentioned the term "encryption" a number of times already. Encryption is a very effective security technique gaining greater acceptance in the networking and security worlds. The reason is that encryption covers so many security exposures with one technique.

Encryption takes data to be protected, along with a "key," and applies a series of mathematical transformations using the two. What results is an encrypted version of the original data. When you encrypt data like a network message, it is changed in such a way that the original contents cannot be recovered without knowledge of the key. As long as the encryption algorithm is of good quality, and the decryption key is kept secret, it doesn't matter who obtains the data, they cannot recover anything meaningful. Encryption also adds authenticity regarding the source of the data. When an encrypted file can be recovered with a key, the person doing the decryption (that is, the reverse of encryption) knows that only someone with the right key could have produced the encrypted file.

Encryption can be supplemented with a mathematical function known as a hash, checksum, or message digest, which is computed based on the contents of the file or message. The message digest is calculated and included with the data before it is encrypted. After decrypting, the digest can be checked against the contents produced. If the computation yields the same results, the integrity of the message has been verified.

Encryption at the wireless LAN level. Some wireless LAN products offer optional encryption mechanisms that can greatly enhance the security of data passing over the network. For example, a node on a Proxim RangeLAN can include a Digital Encryption Standard (DES) chip that encrypts all data sent out. The key used for the DES algorithm must be set the same for all nodes participating on the LAN. Once set, only authorized nodes (those with the right key) can decipher the content of messages sent across the LAN, or send encrypted messages. With one solution, a variety of security evils are covered. See Figure 9.8 for an illustration of encryption.

This approach to encryption is known as single-key or symmetric encryption because the same key is used to both encrypt and decrypt data. (You'll learn more about a two-key system in the next section.) Single-key encryption algorithms like DES have the advantage of being very fast. However, they suffer the problem of having to distribute the single key to all parties in a secure fashion over a secure channel. In a wireless LAN, this can be accomplished during the initial setup. On the other hand, updating all the nodes for a key change would take considerable effort in a larger wireless LAN, so it's not likely to be done very often.

If this technique is used, all traffic on the wireless LAN is encrypted. This involves some performance penalty (although using the DES algorithm in hardware usually provides fast encryption). As an alternative, encryption can be applied within specific applications in order to protect data before it is even sent on the network, as you'll see in the next section.

Encryption at other levels. Rather than encrypting all traffic between all nodes, two nodes can agree to use encryption for only certain messages. In this way, they can very effectively protect sensitive traffic without incurring the performance penalties of encrypting data that doesn't really need it. Once again, if encryption is used, it won't matter who taps into the wireless LAN. The data remains safe.

A good example of this type of agreement can be seen using a common network application: electronic mail ("e-mail"). Two correspondents can agree on a single key to use, then encrypt all their messages before sending

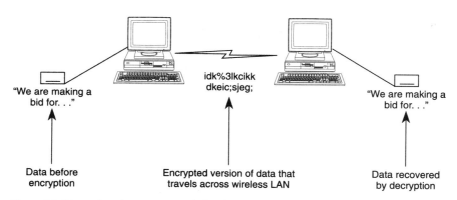

"We are making a
bid for. . ."

idk%3lkcikk
dkeic;sjeg;

"We are making a
bid for. . ."

Data before
encryption

Encrypted version of data that
travels across wireless LAN

Data recovered
by decryption

Figure 9.8 Encrypting data sent on a wireless LAN.

them using the e-mail system. The mail system, the network software, and the wireless LAN hardware would all be unaware that encrypted traffic is being sent. And any network attackers would be foiled. But the correspondents have to get the key to each other somehow.

A somewhat different approach to encryption, using a two-key system, is gaining popularity for use with e-mail. Under the two-key or asymmetric method, one key is used to encrypt data, while a second key (which cannot be derived from the first) is used to decrypt the message.

When applied to e-mail, each correspondent has both a private key and a public key. As you might have guessed, the private key is kept secret, but the public key is distributed to everyone and anyone who wants it. If you encrypt data with someone's public key, only that person can decrypt it with his or her private key. Conversely, if a message can be decrypted with someone's public key, then that person must have been the one to encrypt the message with his or her private key.

Using public and private keys in this way yields four main functions, which are very similar to sending regular mail (only much more secure):

- Signing a message in order to prove you created it. This involves computing a message digest, then encrypting the digest with your private key.

- Sealing a message in order to protect its confidentiality. This requires encrypting the message with your correspondent's public key.

- Opening a message so that the contents can be viewed. Only the intended correspondent can decrypt the message, using the private key.

- Verifying a message to ensure that it has come from who you think sent it. Using the public key of the sender is the only way to recover the message digest and check the contents.

When the preceding steps are in place, as shown in Figure 9.9, the recipient can be assured that the message has come from the signer, that the message has not been changed in any way, and that the recipient is the only person who could recover the message. This approach means that a message could go over any means, whether secure or not, yet the confidentiality and integrity goals will still be met.

Another advantage with two-key encryption is for key management. Various schemes are possible, including central certification agencies, to distribute a person's public key so that a recipient can reliably authenticate a message. In addition, you can publish your public key just as you do your telephone number. The problem of keeping a single key secret while communicating it between parties is greatly reduced. Using a technique like encryption for sensitive data can reduce the pressure for extensive security mechanisms and procedures over other wireless LAN components.

So, we'll try once more. "Are wireless LANs secure?" The answer is that they can be made secure if the right combination of security measures are in place. The specific ones you use will depend on your circumstances and business needs. In the next chapter, you'll learn how to manage your secure wireless local area network.

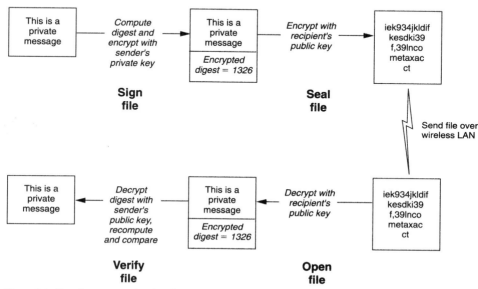

Figure 9.9 Two-key encryption for electronic mail.

10

Implementing and Managing Wireless LANs

You've decided that wireless LANs have a place in your organization. You've formed a clear idea of how you are going to use and benefit from the technology. Hopefully, you've done some preliminary cost-versus-benefit analysis, all the better to make a convincing business case for the expenditures you face. All that's left now is to buy the equipment and install it, right?

We know you're not that easily fooled. If you've ever been involved in implementing a new technology, you know that you're just beginning. Pressing ahead without planning how to implement your wireless LAN, or considering how to support it in the future, is a recipe for disaster. Murphy's Law is, "If anything can go wrong, it will," and this has perfect applicability to installing wireless LANs.

There's ample opportunity for Murphy's Law and its corollaries to function during the implementation and ongoing support of your LAN, without tempting fate by not planning. Installing a wireless LAN in your organization, and keeping it going, will be much simpler and hassle free if you follow the suggestions in the next sections. They describe important steps to follow when implementing a new wireless LAN, plus key considerations that need to be addressed in running the LAN on a day-to-day basis.

Implementation

Getting into a new technology involves risks to your organization, its information assets, and perhaps your career. A massive expenditure of time and money is dangerous, unless you take steps to diminish the risk. So you have to hedge your bets, either by limiting the amount you can lose, or by taking steps to make it a sure thing.

Pilot projects

One way to limit the amount at risk is by creating a pilot project. A pilot works best when you can select a small test site in the organization that has the same computing network requirements as the larger user community. The idea is to set up a limited-size, limited-cost version of the system that can offer a proof of concept, without the expense of a full implementation.

Setting up a pilot wireless LAN provides extremely valuable real-world experience. A pilot will quickly point out any gaps between the theoretical benefits of a wireless LAN and the practical pitfalls. It also can help identify advantages or applications not originally considered. Using a pilot project delivers an understanding of how to use wireless LANs that simply cannot be obtained from sales brochures or even from a book. A pilot also limits the risk to your information assets.

Selecting a pilot site. For wireless LANs, a good choice is a small group of users that regularly work together. Perhaps they already use a wired LAN, but they could benefit from being untethered. Naturally, the group must make use of LANs in a similar fashion to others in the organization. Otherwise the experience gained in using the wireless LAN will not apply to other groups in the company. For example, in a sales office consisting of many teams, one team made up of managers, sales staff, and assistants can serve as the pilot user group.

Pilot project planning. Even though the pilot involves a smaller LAN, the expenditures involved still can be high enough that the technology isn't exactly "throwaway." Setting up the pilot site also will require time from staff and users to install equipment, make adjustments to network drivers and applications, and deal with problems. That time carries a cost to it. Therefore, a business case for the pilot project still should be developed. Fortunately, the smaller expense for the project should make it easier to justify, especially when compared to a full-blown implementation. You should also construct a plan for setting up the project. (You see more about business cases and planning later in the chapter.)

Installation. Working with the users, choose and install an appropriate wireless LAN. Depending on your company's purchasing power, you might even obtain the necessary equipment on a trial basis. (This also will depend on the demand for wireless products.)

Make sure the installation will be safe, with adequate provision for the inevitable problems. For example, an existing wired LAN could be left as a fallback, in the event that the wireless LAN proves unusable or unreliable. Also, plan to install the LAN at a time that will not disrupt the work of your users (or at least not much).

Testing and monitoring.   Once the wireless LAN pilot is up and running, it's important to closely monitor the project and obtain feedback from users about their experiences. Throwing the technology in with a "sink-or-swim" attitude will almost certainly be unsuccessful. If there is no support or remedial action when problems occur, users quickly will abandon the wireless LAN, unless they are highly motivated to keep things running themselves.

It helps if the users are technically sophisticated, at the level of "power users." This does have a disadvantage because it limits how representative the pilot findings are. This is because power users might cause more trouble on the wireless LAN than less sophisticated users, or they might deal with problems that others wouldn't tolerate. However, the benefit is that sophisticated users can help you get over the initial hump of implementing a new technology, and problems always will occur. It helps to have some extra resources and expertise around when you are troubleshooting wireless LAN problems. And to specifically test user friendliness, as well as how robust the LAN is, you always can run a second pilot for a short period of time, with less sophisticated users.

During the pilot phase, as much testing as possible should be done. This testing includes at least:

- Trying different physical configurations for the wireless LAN, such as different locations for access points, or for reflectors in an infrared LAN.

- Running the wireless LAN with varying loads (number of users and volume of data transfer).

- Simulating peak usage times (for example, when everyone tries to log on at the same time in the morning).

- Using a full array of network-based application software to determine if each one suffers through use on a wireless LAN.

The wireless LAN also should be subject to difficult environments, such as where there might be a large amount of electrical noise and interference.

Eventually, based on the findings of the pilot, your organization might decide to implement wireless LANs on a larger scale. You still will need to follow all of the steps described in the next section for your full implementation. But your experience from the pilot project should make tasks like building the larger business case and planning the implementation much easier to do.

Performing the full implementation

Using a pilot project helps to diminish the risk of adopting new technology by limiting the investment made and reducing the amount at stake. A further way to reduce the risk is through proper planning for a full implementation. Proper planning leads you through a thought process that helps to identify and correct problems before they happen. It also allows you to examine the full scope of costs and benefits in the cold light of day. Planning doesn't guarantee a successful implementation, but it greatly decreases the chance of failure. And when something does go wrong, at least you followed the prudent business steps. These processes will help mitigate the damage to the project (or at least to your job).

Your circumstances might force you to move directly to a full implementation, bypassing a pilot phase. For example, an imminent office move, competitive pressures, or the urgent need for a new wireless application might force you to install a wireless LAN more quickly than you would normally. Resist the temptation to avoid planning your implementation in an effort to save time. If you plan well now, you'll avoid problems in the future. Remember the old saying, "There's never enough time to do it right in the first place, but there always seems to be time to do it over." Or, "If you fail to plan, you plan to fail."

Building a business case. As mentioned earlier, whether or not you use a pilot project, you likely will face building a business case to justify full implementation. Without the experience gained from the pilot project, you'll have to use more assumptions and predictions, and you'll be less certain of the results. Try to think of all the costs you'll incur as part of the project, including the price of wireless modules for workstations, central access points, and bridge points. If you're using a microwave LAN that operates on a dedicated frequency, then you'll have to include the costs of licensing (government fees, site survey, preparation and inspection), which will probably involve use of outside expertise in the field of data communications. And of course, there's the time required on the part of technical and user staff to install and configure the wireless LAN.

In choosing a wireless LAN, it might be tempting to focus on the solution available from one vendor. Especially with the unique and varied functionality offered by vendors today, comparing products is more difficult. However, with a little extra effort, the capabilities of products can be normalized, allowing for a comparison based on price and performance. In some cases, the products are identical, just offered by competing companies. For example, WaveLAN, produced by AT&T Global Information Systems, is also marketed by Digital Equipment Corporation. So it might still be possible to get a better price by clearly defining your requirements and purchase selection criteria, then soliciting competitive bids from suppliers.

Planning the implementation. An implementation plan is crucial to the success of the project. The plan should include a detailed schedule and should address the specific tasks required to get the wireless LAN up and running. In constructing the plan, you should consider areas such as:

- Installation of wireless nodes. How involved an installation is depends on the wireless LAN you select. It simply might be a case of unplugging a LAN cable from the wall and inserting it into the wireless node. Or it might mean opening up each and every computer to insert a new wireless NIC. Your organization might require many different kinds of NICs. For example, in the PC environment, you might have ISA, EISA, MCA, or PCMCIA cards. To further complicate matters, you also might have AppleTalk or EtherTalk cards. And if your organization uses many different computers, software packages, and configurations, conflicts with the new NIC will invariably occur. That means troubleshooting problems, changing jumpers, moving other cards, and reconfiguring software. Some installations will go very smoothly, while others only go kicking and screaming. So make sure you allow time in your implementation plan for problems in installing nodes.

- Installing access points and bridges. Again, the time required to perform this task depends on the type of wireless LAN you selected. Because wireless access points and bridges have such an impact on the rest of the LAN, it's important that they are placed (or sited) to provide optimum performance and coverage. Count on spending some time reconfiguring them for best results. Where a bridge product uses dedicated microwave frequencies, expect time for site preparation and inspection so that you comply with all applicable licensing regulations.

- User training. Fortunately, many wireless LAN products are, for all intents and purposes, "plug and play." They almost are transparent to the user, so if they already know how to use a LAN, then converting to a wireless LAN shouldn't require additional training. However, the use of wireless technology also can bring about some operational changes. A short and simple training session, or tutorial document, can be a good forum to bring these up. For example, file transfers, or even initial login to a file server, might be a bit slower than a person is used to. They should be cautioned against immediately assuming there are problems (usually followed by repeatedly hitting the Escape key or pressing the Enter key harder). Users also must be told about the possibility of interfering with the wireless LAN, such as by inadvertently blocking an infrared transceiver. The training session also can be a good time to deal with any concerns users might have about health and safety issues.

- Testing and problem resolution. Each installation should include adequate testing procedures to ensure success. The person performing the installation should at least ensure basic functionality before turning the equipment over to the user. The user can then perform more extensive tests using their normal LAN-based applications. Guidance should be given to users on the types of tests they should perform, and what to do when things go wrong. You also must allocate time and resources to problem resolution after the initial installation.

- Information security. Your implementation plan should include an evaluation of the security requirements for each new wireless LAN. Before the LAN is installed, adequate security measures should be identified and put into place. This helps avoid security problems during the installation phase when new users are most vulnerable. A security breach entrenched in the system at this time might be impossible to remove later.

- Updating the help desk. Where your organization provides LAN users with a help desk function, the help desk staff must be trained in using wireless LANs and recognizing problems that might be caused by the technology.

- Provision for spare components. Even though you are installing new equipment, there is a chance that the installation and testing process will identify faults. It's useful to have a readily available source of some spare wireless components, particularly nodes, which can help keep the implementation on schedule.

- Contingency plans for fall back. Problems can occur, even with the best laid plans, and even after many successful implementations. For example, placing a wireless LAN in a new area might fail due to a previously unknown source of interference. For each installation, there should be a fallback plan in case the installation fails. The easiest case is when a wired LAN is being replaced by wireless. Users can simply return to the cable. However, for a new installation, more creative solutions will be necessary.

Implementing the plan. Once planning is complete, the real process of rolling out the equipment can begin. As a practical matter, you probably will implement your wireless LANs in phases, installing new equipment based on logical work groups or coverage zones. After each installation, it's a good idea to review any problems that occurred and revise plans for future roll-outs.

As use of the wireless LAN takes hold in the company, and the planned benefits start to accrue, you can reflect on the success of your planning and implementation efforts. But wait. Now you have to provide ongoing support. (You knew there had to be a catch to this.) You'll see more on managing your wireless LAN in the next section.

Ongoing Management

Getting your wireless LANs up and running might seem like quite a challenge in itself. And it is, as is working with any new and different technology. Keeping a strange new technology operational is no less of a challenge. If you have experience in running a wired LAN, then it will serve you well when dealing with wireless. However, with a more complex technology involved, compared to simple cabled LANs, you might run into new or unique problems with wireless LANs. And you might have to seek outside help on more occasions. The following sections examine the areas in network management that are affected by using wireless LANs.

Fault management

Fault management is the process of dealing with network failures. Ideally, emphasis is placed on preventing failures from happening, rather than correcting those that do. This requires proactive fault management techniques designed to prevent a fault from turning into a failure. These include:

- Avoiding faults in the first place.

- Masking or limiting the effect of a fault that does occur.

- Compensating for a fault that does occur.

Of course, there are times when those proactive techniques cannot prevent a full-blown failure from happening. Then it is necessary to use reactive fault management techniques, which basically boil down to identifying and fixing the fault.

Wireless LANs might seem to have an immediate advantage over wired LANs because there are no cables to cause problems. LAN cables are often subject to abuse, particularly at connection points. For example, the repeated attaching and removal of a notebook computer to a wired LAN can stress the connectors so that they eventually break. You also might damage cables along their length, such as while being pulled through conduit or run over at a temporary installation. Eliminate the cable, and you've avoided one source of faults.

Unfortunately, the increased complexity of wireless nodes introduces additional sources of problems. Conventional wisdom taken from the farm expresses this well: "The more riggin', the more friggin'." For example, the additional circuitry required to produce radio frequency signals and apply data through modulation can be a further source of problems.

The sometimes mysterious world of wireless communications also can be a source of problems. A wireless LAN that works perfectly well one day might all of a sudden not work at all the next. All it takes is a new source of interference appearing in the area.

Because wireless LANs face a more hostile environment, they feature more techniques that mask faults. The best example is use of spread-spectrum techniques for transmission, which are more resilient to interference, both from other wireless LANs and from other forms of electromagnetic radiation. Wireless nodes also are shielded very well to prevent interference from outside sources.

The hostile wireless LAN environment also creates the need for error-correcting data codes and collision avoidance contention techniques, both of which help compensate for naturally occurring faults like electrical noise.

What becomes more difficult with wireless LANs is identifying the fault. In a cabled LAN, it usually is possible to isolate the cause of a problem by selectively disconnecting portions of the LAN. When the problem stops, you've found the culprit. These techniques, plus some simple test instruments like an ohmmeter or continuity tester, often are all you need to identify where a problem lies.

With wireless LANs, however, it can be far more difficult to isolate the problem area. For example, isolating portions of the LAN from others might require physically moving it to a new location that is shielded or sufficiently distant from the rest of the LAN. Another example is interference, which might be intermittent. Problems that only occur once in a while require that the LAN be monitored much longer before a problem is found. There also can be more trial-and-error activities when configuring a wireless LAN, such as when physically moving or reorienting components to try to get reliable communication. Contrast this with a wired LAN, where a node is connected or it isn't.

Finally, the test equipment required to monitor and analyze radio signals, or test signal strength, are far more complicated and expensive. Problem diagnosis for a wireless LAN can still be done in-house, but only some faults can be identified and fixed. You probably will need to use an outside consultant to tackle the more challenging difficulties.

Advances in diagnostic software, which is built in to network devices and available to network managers, will help identify and correct faults in both wired and wireless networks. The availability and quality of software aids should definitely be a consideration when evaluating wireless LAN products.

Performance management

Monitoring the performance of a LAN is a proactive technique for identifying where future problems might occur. In this case, "problems" mainly relate to user complaints about how quickly they can get information across the LAN.

Managing the performance of a wireless LAN is not dissimilar to wired LANs. For example, network traffic can be analyzed to identify bottlenecks. Heavily used LAN segments can be partitioned with a bridge. Focal points, such as a network backbone, can be switched to faster technology. Using different components and technologies, all of these techniques are available to an organization with wireless LANs. In some instances, the techniques are easier to apply because they are wireless. For example, adding an extra access point can easily be done without rerouting cables.

However, the performance of a wireless LAN also can be affected by parameters that are unique to the technology. Recall the discussion from chapter 3 on spread-spectrum techniques. Judicious selection of spreading codes for DSSS or FHSS can reduce interference and improve network throughput. There also is a parameter such as power output. Decreasing the power of one wireless device might sometimes be as beneficial to performance as increasing the power of another.

The problem is that modifying these values normally requires a good deal of experience in tuning the performance of a wireless link. Normally, it is not something that a network manager would want to attempt. You might not even be able to modify these parameters even if you want to. Rules governing use of the radio spectrum often prohibit the modification of any operational aspect of equipment.

Even when a parameter can be legally adjusted, it might not be accessible to you. The complement of the ease of installation for "plug-and-play" systems is the difficulty of making changes to how they operate. It's like a television that automatically adjusts itself for tint and color intensity (as most do). If all works as it should, there are no concerns. But when the picture develops an odd hue, and you find that there is no manual means of making adjustments, you can become very frustrated.

Similar to fault management, advancements in performance management software will present key operating parameters to the network manager, along with suggested changes. Facilities such as these also should form part of your evaluation criteria for wireless LAN products.

In the meantime, if you believe you have a need to do fine tuning on operational parameters for a wireless LAN, then you would be well-advised to consult a communications engineer or other expert in the field. They can bring the right equipment to bear on the problem and help identify safe, reliable, and legal solutions.

Asset management

Chapter 9 stressed the importance of maintaining an inventory of wireless LAN components. This is a key step to protecting the company's assets, particularly mobile assets like wireless LANs. Also, it might be necessary to keep an inventory when you need to allocate the cost of components to each user department, or to charge directly for equipment purchased on the department's behalf.

Here again, future software products might provide great assistance such as automatically recording devices that appear on the wireless LAN and noting those that have not been heard from for a while. Aids like these should also be a consideration when evaluating wireless LAN products. Speaking of the future, the next chapter discusses what's on the horizon for wireless LAN products.

The Future

"My interest is in the future because I am going to spend the rest of my life there."

Charles F. Kettering

"I never think of the future. It comes soon enough."

Albert Einstein

"The only limit to our realization of tomorrow will be our doubts of today. Let us move forward with strong and active faith."

Franklin Delano Roosevelt

11

Future Wireless Networks

As history shows us, technology creates its own previously undreamed-of uses. Alfred Bernhard Nobel thought his invention (dynamite) only would be used for good. After inventing dynamite, he is quoted as saying, "My dynamite will sooner lead to peace than a thousand world conventions. As soon as men will find that in one instant whole armies can be utterly destroyed, they surely will abide by golden peace."

Obviously, Nobel was absolutely and undeniably wrong. It is impossible to predict the uses of innovative technology, and you cannot even try to predict the uses for innovations until you have the technology. Thorstein Veblen made a play on Jonathan Swift's famous phrase, once saying, "Invention is the mother of necessity."

You must look at today's emerging technologies and try to predict how they can help to transform your business. Again, history shows that, in many situations, supply creates its own demand. What does this mean? Well, people don't know they want something until they see that they can have it. Then they feel they cannot live without it! A good example of this phenomenon is television. When they first introduced television, developers and proponents never envisioned today's popularity.

You can find other examples in business, in say, photocopiers. When Xerox first introduced the photocopier, most people saw little use for it, but the technology created its own uses. People started making copies of everything. Photocopiers did more than just replace carbon paper and mimeographs, its supposed purpose. More than just a replacement technology, photocopiers created a whole new market of convenience copies.

The Failure of Wireless Technology

Therefore you cannot simply ask people how they would use a technology like wireless LANs in their business. Invariably, they will reply with a description of how the technology might improve a task they already do, such as replacing cable in an existing network. But wireless local area network technology is a breakthrough technology. Breakthrough technologies make feasible possibilities that people have not yet dreamed. The challenge that most organizations fail to meet is recognizing the business possibilities in new technology. This failure on the part of most corporations is understandable.

The failure with wireless technology lies in not addressing the real question of whether wireless LANs are a viable networking solution. For wireless local area networking to truly succeed, it must progress beyond a cable replacement and create its own niche. Even though cable replacement has cost-saving value, the true value of wireless local area networking will be in transforming how work is done, not in saving money. In short, the value of wireless LANs is that it will allow companies to do something they had not previously done.

To summarize, the real power of technology is not that it can make the old processes work better, but that it enables organizations to discard old processes or rules and create new ways of working. In truth, you should find the long-standing rules that technology will allow your organization to break, then see what business opportunities you can create by breaking those rules. Wireless LANs break the rule that workstations or network devices must be physically attached to the network. Now it is possible for people to be located anywhere in the office and communicate directly with the system.

That insight gives a company a powerful tool for transforming its operations. Look at how hand-held communicators changed the operation of a restaurant. Wireless LANs enable restaurants to increase throughput, and hence profits.

The Promise of Wireless Technology

With wireless data communications, field people can request, view, review, manipulate, use, and transmit data almost anywhere without ever having to run back to the office. With increasingly miniaturized input devices, people can connect to information sources wherever they are. Take, for instance, rental cars. When a car returns to the lot, an attendant, equipped with a tiny hand-held communicator, meets the car, pulls up the record of the rental transaction, and enters the mileage and calculates the charge. Customers are elated because they never have to visit the office. The rental company is likewise elated because input information is more accurate, and they can clear customers faster.

Let's recap the current state of wireless technology and understand how it enabled these new applications. In chapter 1, you learned that wireless transmission technology for premises applications is still in an introductory stage, with UHF and microwave radio having attracted the most attention so far. The primary goal of this technology is as a substitute for premises cabling, although an untethered workplace is now a growing motivation. An untethered approach is supported by the industry's trend toward miniaturization manifested in smaller portable personal computers such as notebooks, laptops, and palmtops.

As a premises wiring substitute, you have seen that cordless telephones, PBX adjuncts, campus links, and 2-Mbps LAN systems are already available in the marketplace. As mentioned, there are some technical hurdles for radio in supporting useful distances, data rates, and flexibility, but these currently available systems demonstrate the feasibility within a range of applications. It is interesting to remember that even with significant regulatory restrictions, radio is being vigorously applied and marketed. This alone provides some glimpse of its potential value to organizations exploring this technology.

Another wireless technology, line-of-sight infrared transmission, has been applied to telephone headsets, campus telecommunication, and now LAN equipment. So far, market response to the 16-Mbps infrared Token Ring throughput systems that are currently available has been limited. This equipment requires optical transceivers that must be precisely positioned, creating situations that can sometimes detract from office aesthetics. Campus links for telephone and data systems have been available and successfully used for many years.

Proprietary radio systems using general and licensed frequency bands already are on the market. A close examination of the limitations of radio has led most analysts to conclude that it will not substantially eliminate customer-premises cabling, but it will become mostly an extension option for existing cabling systems.

The regulatory situation provides a major impediment. Frequency allocations for premises radio are still not globally settled, and support of 10 and 16-Mbps LANs require very large allocations of radio spectrum. Accommodating a 100-Mbps LAN will be significantly more difficult for both allocation and technical solution. Already, regulators and developers have recognized the need for a wide block of radio spectrum, and committees are working to establish radio LAN standards.

In spite of the demand for allocations and standards, the difficulties in establishing these will result in a delay of a couple of years for a standards-based LAN. Once established, standards-based equipment will improve the radio LAN market segment. In this time frame, however, 10- or 16-Mbps radio LAN equipment will then be one of many extension options for the newer switching technologies.

Currently available wireless systems do not substantially eliminate cabling, but they rely on cabling for system interconnection and use the wireless to accommodate user portability and flexibility. One plausible scenario of the future is that telephone and palmtop computing and multimedia will evolve back into a more centralized architecture and encompass local and wide area networking.

The true value for wireless local area networking will come when it melds with wireless wide area networking. In the not so distant future, say by the year 2010, wireless data communications will be ubiquitous. Notebooks, subnotebooks, laptops, palmtops, hand-held communicators, and personal digital assistants all beg for wireless data communications. Conceivably, you will possess a truly hand-held communicator resembling today's personal digital assistant from Apple and Sharp. Your personal digital assistant will have built-in wireless capabilities that will allow you to remain in constant communication with your business regardless of your location on the planet. Like its predecessor technology itself, wireless data communications will change the way people work. You will no longer need to go to work to have current and accurate corporate data. Increasingly, people will use wireless communications to access people and information anytime, anywhere, and from any place.

It should become clear from the examples in this book that further advances in technology will break more rules about how you conduct business. Rules appearing inviolate today might become obsolete in a year or less as the technology evolves. For instance, 100-Mbps technology at an economical price might spur the wireless marketplace to action. Consequently, your evaluation of wireless LAN technology is not a one-time exercise. It should not stop with this book if you want to dramatically move ahead of your competitors. On the contrary, you should stay on top of this emerging technology and investigate how it can transform your business.

It takes patience and business acumen to figure out how this new technology applies in your situation, or to see beyond the obvious cable replacement applications to novel applications. At first blush, wireless LAN technology looks as if it will only marginally improve your business. However, look at how some of the companies mentioned in this book have transformed their business and start thinking on a daily basis about yours.

Think Creatively

How do you go about looking for breakthrough applications? Well, the best way is to practice the five stages of creative thinking:

- Initiation. Obtain all the facts about your organization and the problem. This might involve restating or redefining the problem. This step usually involves convergent or analytical thinking.

- Industriousness. Dream up possible solutions through either brainstorming, JAD (joint application development) sessions, synetics, attributes listing, or lateral thinking. This step typically involves divergent or cognitive thinking.

- Incubation. Let the ideas and problem sit in your subconscious to give you time to accept the ideas or to form new ones.

- Insight. Recognize possible solutions to the problem. This step also involves divergent or cognitive thinking.

- Induction. Analyze all the ideas from the last three steps for possible solutions. This step also involves convergent or analytical thinking.

When you practice the five creative-thinking steps, those breakthrough applications eventually will come. You cannot read about this new technology today and roll it out tomorrow. It takes time to study it, to understand its significance, and to plan its rollout. Inspiration involves perspiration. Remember. Roam wasn't built in a day! Hopefully, this book has started you thinking about how to use wireless LANs in your business. Good luck and inspiration.

Appendices

"Any sufficiently advanced technology is indistinguishable from magic."

Arthur C. Clarke

"It is difficult to say what is impossible, for the dream of yesterday is the hope of today and the reality of tomorrow."

Robert H. Goddard

"He that will not apply new remedies must accept new evils; for time is the greatest innovator."

Francis Bacon

The Vendors

There are a number of vendors offering wireless bridges, hubs, and network interface cards. The following is a list of current products. Please note that there has been no verification of claims by vendors, and information provided, though adapted, is from the vendors.

AIRONET Wireless Communications, Inc. (formerly Telxon Corporation)

3330 West Market Street
P.O. Box 5292
Akron, OH 44334-0292
Voice: Toll free: 1-800-800-8016
Direct dial: 1-216-873-2000
Fax: 1-216-873-2910

Product

ARLAN products

Product overview

The ARLAN product line uses spread-spectrum and CDMA. Using a conventional cabling system, these products attach to the cable and provide a full range of interconnectivity options. ARLAN technology uses the concept of a microcell, which is integral to their product line. By using multiple base station antennae, you can extend the network to create microcells, each with its operating systems and devices. The Microcellular Architecture technology handles multiple overlapping cells to create a seamless network within a building, warehouse, or plant. Hand-off from cell to cell is handled by the system, thus allowing for connectivity anywhere the signal can be heard. Also, using spread-spectrum technology, the system can spread the information across various frequencies with different coding techniques. Therefore, multiple devices operating within the same area serving different needs can co-exist.

Figure A.1 shows the spread-spectrum radio module used by the ARLAN products and can be integrated by Original Equipment Manufacturers into their products. A vendor can integrate the module into a portable or fixed station product to provide wireless communications. AIRONET has an installed base of more than 200,000 wireless nodes (more than 50,000 are spread-spectrum) at more than 5,000 locations.

Product

ARLAN 610 and 610E Wireless Ethernet Hub

Figure A.1   ARLAN's spread-spectrum radio module.

Product overview

The ARLAN 610 wireless hub provides connectivity between cabled and remote devices equipped with either an ARLAN 655 ISA bus, ARLAN Micro Channel, or other ARLAN 610 nodes performing concentrator node functions. The hub uses spread-spectrum technology and can be used anywhere on an Ethernet or Token Ring LAN.

Specifications

Technology: Spread-spectrum
Range: 610: Up to 3000 feet indoors, up to 6 miles outdoors (line-of-sight);
      610E: Up to 1500 feet indoors
Frequency: 610: 902–928 MHz (ISM band, no site licensing required), 610E:
      2.460 GHz
Data rate: 610: Up to 1.35 Mbps; 610E: 230 Kbps to 1 Mbps
Compatibility: Protocol independent

Product

ARLAN 620 and 620E Wireless Ethernet Bridge

Product overview

The ARLAN 620 and 620E can bridge two Ethernet LANs and make them appear as if they are on the same cable. The devices have a console port interface that can be used locally through an attached terminal or remotely by modem to monitor unit status and performance.

Specifications

Technology: Spread-spectrum
Range: 620: Up to 3000 feet indoors, up to 6 miles outdoors (line-of-sight);
     620E: Up to 1500 feet indoors, up to 3 miles outdoors (line-of-sight)
Frequency: 620: 902–928 MHz; 620E: 2.460 GHz (ISM band, no site licensing required)
     ing required)
Data rate: 620: Up to 1.35 Mbps; 620E: 230 Kbps to 1 Mbps
Compatibility: Protocol independent

Product

ARLAN 655, 670 and 690 Wireless Network Interface Cards

Product overview

The ARLAN product line includes network interface cards and LAN adapters. The ARLAN 655 is a half-size network adapter card installable in any Industry

Standard Architecture (ISA) bus computer. The card comes equipped with a communications processor and a spread-spectrum radio for high-speed wireless communications. Figure A.2 shows what the ISA card looks like. Notice the radio antenna, which acts as a transceiver.

The ARLAN 670 is a full-size network adapter card installable in any Micro Channel Standard Architecture (ISA) bus computer. As with the ARLAN 655 card, the 670 card comes equipped with a communications processor and a spread-spectrum radio for high-speed wireless communications. Figure A.3 shows what the micro channel card looks like.

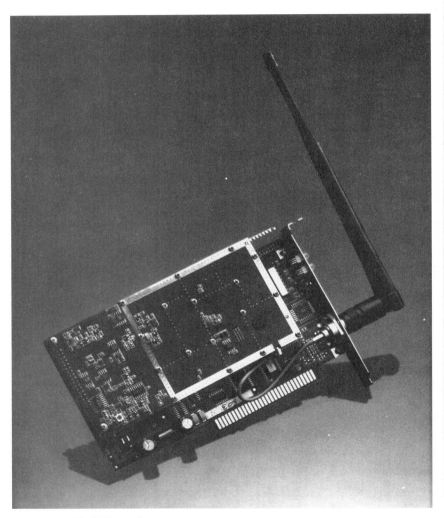

Figure A.2 ARLAN ISA card.

The recently announced ARLAN 690 is a PCMCIA network adapter card installable in any computer with a PCMCIA slot. As with the ARLAN 655 and 670 cards, the 690 card comes equipped with an attached communications processor and a spread-spectrum radio for high-speed wireless communications. Figure A.4 shows what the PCMCIA device looks like. These cards can communicate with either another computer with an ARLAN wireless network adapter or an ARLAN 610 Wireless Ethernet hub.

Specifications

Technology: Spread-spectrum

Range: 655, 670 and 690: Up to 3000 feet indoors, up to 6 miles outdoors (line-of-sight); 655E, 670E and 690E: Up to 1500 feet indoors, up to 3 miles outdoors (line-of-sight)

Frequency: 655, 670 and 690: 902–928 MHz; 655E, 670E and 690E: 2.4 GHz (ISM band, no site licensing required)

Data rate: 655, 670 and 690: 215 Kbps to 880 Kbps; 655E, 670E and 690: 230 Kbps to 1 Mbps

Compatibility: Protocol independent

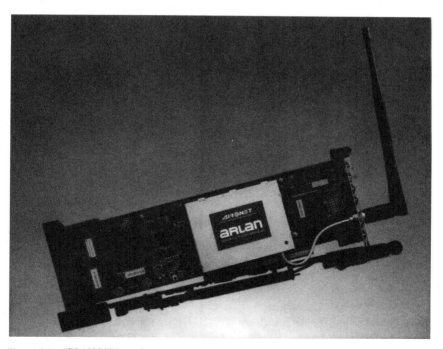

Figure A.3 ARLAN MCA card.

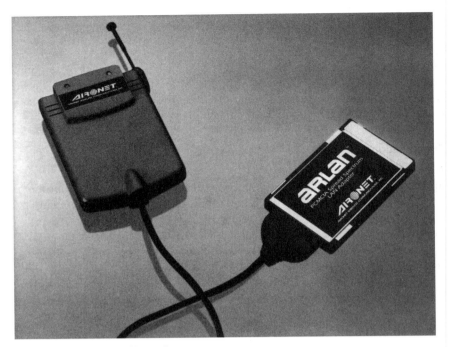

Figure A.4 ARLAN PCMCIA card.

Product

ARLAN 680 Wireless Network Adapter

Product overview

The ARLAN 680 is an outboard transceiver that connects to any computer with a parallel port. The device comes equipped with a communications processor and a spread-spectrum radio for high-speed wireless communications. Figure A.5 shows what the LAN adapter looks like. Again, notice the radio antenna, which acts as a transceiver. This device can communicate with either another computer with an ARLAN wireless network adapter or an ARLAN 610 Wireless Ethernet hub.

Specifications

Technology: Spread-spectrum
Range: 680: Up to 3000 feet indoors; 680E: Up to 1500 feet
Frequency: 680: 902–928 MHz 680E: 2.460 GHz (ISM band, no site licensing
          required)

Data rate: 680: Up to 1.35 Mbps; 680E: 230 Kbps to 1 Mbps
Compatibility: Protocol independent

Alps Electric (USA), Inc.

3553 North First Street
San Jose, CA 95134
Voice: Toll free: 1-800-825-2577
Direct dial: 1-408-432-6000
Fax: 1-800-825-1445

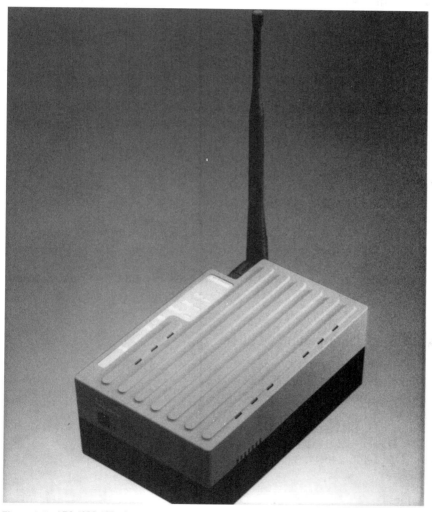

Figure A.5   ARLAN LAN adapter.

Product

RadioPort Plus

Product overview

ALPS offers a RadioPort Plus wireless LAN adapter, which attaches externally to the parallel port of any computer. The RadioPort is a spread-spectrum device offering a range up to 800 feet and an over-the-air data transmission rate of 242 Kbps. To reduce the risk of interference from other RadioPort devices, RadioPort offers three separate transmission channels.

Unlike infrared wireless LAN adpaters, RadioPort's spread-spectrum technology does not require line-of-sight transmission. RadioPort will transmit through walls, ceilings, and floors to make a connection. Though such obstructions might reduce the maximum range, RadioPort still has an effective range of several hundred feet in a typical office setting.

RadioPort wireless adapters are especially effective for low-volume transmission tasks such as electroinic mail, printer sharing, and other applications that do not call for a high-bandwidth network.

Specifications

Technology: Spread-spectrum
Range: Up to 800 feet
Frequency: 902–928 MHz (ISM band, no site licensing required)
Data rate: 242 Kbps
Protocol: CSMA/CA
Compatibility: Artisoft LANtastic, Novell NetWare Lite, NetWare 2.x and
        3.x, MS-Windows for Workgroups

AT&T Global Information Solutions

(Now includes NCR)
WWIS Network Systems Solutions
1700 S. Patterson Blvd.
PCD/6
Dayton, OH 45479-0001
Voice: Direct dial: 1-513-445-7197
Fax: 1-513-445-2468

Product

WaveLAN and WavePoint

Product overview

This incarnation of WaveLAN is the repackaging of NCR's WaveLAN, per-haps the granddaddy of the wireless products. AT&T also markets a WavePoint access product, which acts like a hub.

WaveLAN adapters include an ISA, Microchannel Architecture, and PCM-CIA cards. The PCMCIA card operates in the 902–928-MHz range, while the larger ISA and MCA cards are available in both 902–928-MHz and the 2.4-GHz band. Each adapter can accommodate a Data Encryption Standard (DES) chip to enhance security. The adapters can be used in combination to form a stand-alone LAN. By adding a WaveLAN card to an existing server that has a wired LAN card, a PC bridge is formed. All WaveLAN users can then gain access to the wired LAN.

As an alternative means of connection, the WavePoint access point can be attached anywhere to an existing LAN. Multiple WavePoints per-mit access throughout a building. However, automated roaming is a future enhancement.

The range of the WaveLAN products varies, depending on the model used. AT&T claims that the PCMCIA card can operate up to 200 feet in a typcial office environment. The 915-MHz ISA and MCA cards are reported to deliver up to 800 feet, while the 2.4-GHz model runs up to 590 feet.

The full size 915-MHz cards offer an interesting option. By using a direc-tional antenna (a "yagi"), two adapters can be used to form the heart of a wireless bridge between two LANs, at a distance of up to 5 miles. The range depends on external factors such as clear line-of-site, as well as the length of cable used between the adapter and the antenna. No license is required to operate the link. The WaveLAN products operate at a claimed data rate of 2 Mbps.

Specifications

Technology: Spread-spectrum (direct sequence)
Range: Up to 800 feet omnidirectional, or up to 8 km directional
Frequency: 902–928-MHz and 2.4-GHz range (ISM band, no site licensing
        required)
Data rate: 2 Mbps
Protocol: CSMA/CA
Compatibility: NDIS, ODI, Artisoft LANtastic, Banyan VINES, IBM LAN
        Server, MS LAN Manager, NetWare Lite, NetWare 2.x, 3.x
        and 4.x, , TCP/IP and Windows for Workgroups

Black Box Corporation

P. O. Box 12800
Pittsburgh, PA 15241

Voice: Toll free: 1-800-355-8001
Fax: Direct: 1-412-746-0746
Toll free: 1-800-355-7999

Product

BestLAN2 Hub and BestLAN2 Bridge

Product overview

The BestLAN2 hub supports Ethernet and is protocol independent. The hub works as a standard node on a hard-wired Ethernet LAN. It has connections for Thin Ethernet and 10Base-T. The supplied omnidirectional antenna provides a range of up to 800 feet indoors. You can add the Short-Range Omnidirectional Antenna to get a range of half-mile, line-of-sight. The BestLAN2 Bridge lets you interconnect your Ethernet networks, up to three miles apart. It is an IEEE 802.1D compliant MAC-layer learning bridge.

Specifications

Suggested pricing (U.S. dollars): BestLAN2 Hub: $6,500, BestLAN2 Bridge:
$7,500
Technology: Spread-spectrum
Range: BestLAN2 Hub: Up to 800 feet; BestLAN2 Bridge: Up to 3 miles
Frequency: 902–928 MHz (ISM band, no site licensing required)
Data rate: 2 Mbps
Protocol: CSMA/CA
Compatibility: Hub and bridge support all Ethernet NOS. Products supported
are ISA, MCA, PCMCIA, Artisoft LANtastic, Novell NetWare
Lite, NetWare 2.x and 3.x, MS-Windows for Workgroups

Product

BestLAN2 Cards

Product overview

BestLAN2 cards are spread-spectrum wireless network interface cards for your laptop or notebook. The card comes with a pocket-sized antenna.

You can set up a workgroup wherever you need one. BestLAN2 lets you move your PCs around as much as you like. You also can add wireless workstations to your wired LAN.

Specifications

Suggested Pricing (U.S. dollars): BestLAN2   PCMCIA   adapter: $1,466,
BestLAN2 ISA adapter: $1,466, BestLAN2
MCA adapter: $1,466

Technology: Spread-spectrum
Range: Up to 800 feet
Frequency: 902–928 MHz (ISM band, no site licensing required)
Data rate: 2 Mbps
Protocol: CSMA/CA
Compatibility: Products supported are ISA, MCA, PCMCIA, Artisoft LANtastic, Novell NetWare Lite, NetWare 2.x and 3.x, MS-Windows for Workgroups

Product

Wireless Print Sharer

Product overview

The Wireless Print Sharer uses infrared to connect personal computers to a shared, remote printer. There is no limit to the number of IBM PCs or compatibles that can share a printer. The infrared beam is 15–20 feet at a distance of 50 feet. It will not travel through walls or closed doors, but it can reflect off walls and ceilings.

Specifications

Suggested Pricing (U.S. dollars): $425
Technology: Infrared
Range: Up to 100 feet
Compatibility: IBM PC or compatible

Cabletron Systems

35 Industrial Way
Rochester, NH 03867-5005
Voice: Toll free: 1-800-355-8001
Direct dial: 1-603-337-2211
Fax: Direct: 1-603-332-9400

California Microwave
Product

FreeLINK/62 wireless Ethernet hub, 260 feet, wireless LAN

Product overview

Not available.

Product

RadioLink

Product overview

RadioLink defines a point-to-multipoint network, with up to 16 devices per transceiver. You can build a cellular architecture where the cells use different spreading codes, but they do not provide the hand-over procedures needed to permit mobile devices to roam throughout the network.

Specifications

Technology: Spread-spectrum (frequency-hopping)
Range: About 450 feet omnidirectional; up to 8 km directional
Frequency: 902–928 MHz and 2.4 GHz (ISM band, no site licensing
        required)
Data rate: 250 Kbps
Protocol: CSMA/CA
Compatibility: RS-232

Communications Research & Development Corporation

7210 Georgetown Rd.
Suite 300/400
Indianapolis, IN 46268
Voice: Direct dial: 1-317-290-9107
Fax: 1-317-291-3093

Product

COMRAD CCL901-DP Wireless Data Link

Product overview

Communications Research & Development markets the COMRAD "wireless data link" products for computer and peripheral equipment applications. These products enable computers to communicate with various computing devices such as other computers, computer-controlled equipment, data acquisition equipment, bar-code readers, laser printers, and plotters, without direct wiring, cables, or expensive networking systems.

COMRAD consists of two transceiver units connected to computers or peripheral devices. Using ultra high frequency (UHF) radio transmissions, operating on the 905–926-MHz radio band, COMRAD provides fast, error-free data transfer up to a range of 200 feet indoors, even where units are separated by walls, floors, or ceilings. In unobstructed areas or outdoors, the range can extend to a third of a mile (1760 feet). COMRAD operates at full duplex allowing simultaneous transmission and reception of data.

According to the vendor, with COMRAD you can:

- Access and transfer files between computers where the database constantly changes or needs to be quickly updated.

- Control or acquire data from remote computers, peripherals, bar-code readers, or laboratory equipment.

- Provide a wireless link from a computer to a wired system.

- Share a printer or plotter between two or more computers.

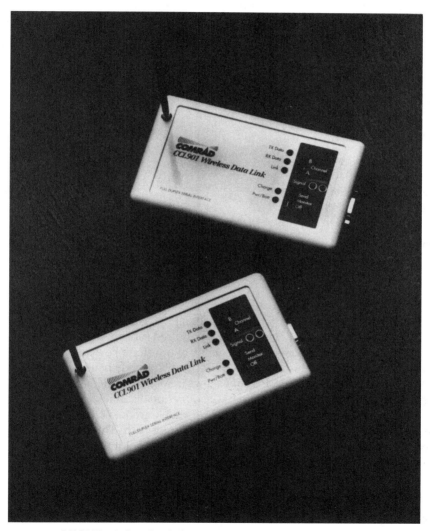

Figure A.6   COMRAD wireless data link.

Each COMRAD package contains both transceiver units, PC compatible software, power adapters, serial cables, and user manuals. Figure A.6 is a photograph of COMRAD Wireless Data Link devices.

Specifications

Suggested pricing (U.S. dollars): Wireless Data Link: $695 (two transceivers and scanner interface); RS-232 serial device: with software $449.95, without software $429.95 (two transceivers)
Technology: Spread-spectrum
Range: Up to 200 feet indoors; up to 1760 feet outdoors
Frequency: Host: 905.050–908.950 MHz, remote: 923.050–926.950 MHz (ISM band, no site licensing required)
Data rate: 38,400 baud
Compatibility: RS-232 asynchronous serial

Cylink Corporation

310 North Mary Avenue
Sunnyvale, CA 94086
Voice: 1-408-735-5800
Fax: 1-408-735-6643

Product

AirLink 64, 64MP, 128, 256

Product overview

AirLink wireless modems offer cost-effective solutions for data and image connectivity, replacing or enhancing conventional wired cabling systems. Spread-spectrum modulation reduces the probability of interference with other radio systems by "spreading" the transmitted energy over a wide bandwidth.

A standard omnidirectional antenna allows indoor or outdoor communication up to 1000 feet. Using an optional yagi directional antenna, you can have line-of-sight links of up to 30 miles. User-selectable frequencies allow multiple AirLink units to operate in the same line of sight path. Three data models support standard synchronous and asynchronous serial interfaces, allowing direct connections to bridges, routers, and communication processors.

Specifications

Technology: Spread-spectrum (direct sequence)
Range: Up to 1000 feet indoors; up to 30 miles outdoors

Frequency: 902–928 MHz (ISM band, no site licensing required); user-
     selectable
Data rate: Up to 64 Kbps (asynchronous) and up to 256 Kbps (synchronous)
Compatibility: V.11, RS-232 or EIA-530

Digital Ocean, Inc.

11206 Thompson Avenue
Lenexa, KS 66219
Voice: Toll free: 1-800-345-FISH
Direct dial: 1-913-888-3380
Fax: 1-913-888-3342

Product

Grouper 100MP

Product overview

The Grouper 100MP is a spread-spectrum, lightweight transceiver to link
the Newton MessagePad to the network. You can send and receive data with
the same security, speed, and reliability as you'd get from an AppleTalk network. Grouper's direct sequencing radio signal penetrates dense office
walls, ceilings, and floors. For maximum security, you can purchase the optional 64-bit DES encryption model.

Specifications

Technology: Spread-spectrum (direct sequence)
Range: Up to 800 feet indoors
Frequency: 902–928 MHz (ISM band, no site licensing required)
Data rate: Up to 2 Mbps
Compatibility: Apple Newton MessagePad, Apple Macintosh, or other
     LocalTalk device using AppleTalk Phase II protocol, such as
     a printer.

Granite Communications

9 Townsend West
Suite 1
Nashua, NH 03063
Voice: 1-603-881-8666
Fax: 1-603-881-4042

Product

VIDEOPAD HI5, VP5 and VP7

Product overview

The HI5, VP5, and VP7 are parts of Granite's VIDEOPAD Links systems. This system consists of wireless, hand-held communicators (VP5 and VP7) transmitting and receiving data from a central base station (BS5) connected to a host computer by an interface card (HI5) at 121 Kbps using spread-spectrum techniques. Figure A.7 shows this arrangement.

The compact BS5 base station controls the flow of data between a VP5 or VP7 and the host computer, and it controls access to the transmission channels. The line-of-sight range for the BS5 is 1000 feet, and for the BS5.5 it is 1700 feet.

The HI5 is the host interface card provided with an RS-422 connector. The VP5 is Granite's hand-held, touchscreen VIDEOPAD using Granite's spread-spectrum VIDEOPAD Links system to provide real-time, two-way wireless communications for service providers. Features include electronic signature capture.

The VP7 is Granite's rugged, hand-held, touchscreen VIDEOPAD featuring electronic signature capture, integral bar code scanning, and Granite's spread-spectrum VIDEOPAD Links system to provide real-time, two-way wireless communications for service providers.

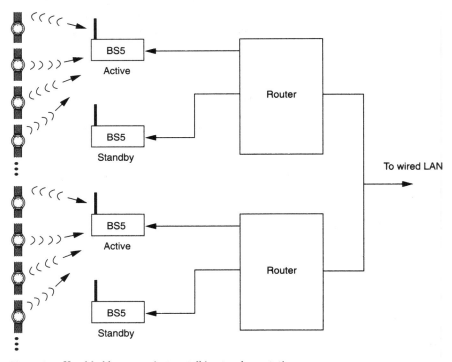

Figure A.7 Hand-held communicators talking to a base station.

Specifications

Technology: Spread-spectrum (direct sequence)
Range: Up to 1700 feet
Frequency: 902–928 MHz (ISM band, no site licensing)
Data rate: Up to 121 Kbps

Hewlett-Packard Corp.

Corvallis Division
1000 NE Circle Boulevard
Corvallis, OR 97330
Voice: Toll free: 1-800-752-0900

Product

Serial Infrared (SIR)

Product overview

Hewlett-Packard Company has a low-cost, wireless, infrared communications interface for communication between personal computer products ranging from hand-held devices to desktop PCs, printers, and other peripheral products.

HP's Serial Infrared (SIR) communications interface is an HP-developed technology that allows wireless, "point-and-shoot" communications between devices positioned within line-of-sight and relatively close proximity to one another. HP SIR provides easy transfer of files and other information between mobile and fixed PC devices and printers, regardless of product brand or size.

HP's SIR data-transfer rate is 115K baud, which allows relatively fast transfer of small files, spreadsheets, calendars, and other data without wires or connection cables. It uses a standard 15660 UART processor and can be implemented in an ASIC chip.

HP pioneered the use of infrared technology in its professional calculators, and more recently in its HP 95LX and HP 100LX palmtop PCs, its HP Omnibook 530 superportable PC, and its HP Vectra 486XM PCs. The HP 100LX can send and receive data at data rates up to 115 Kbps using a built-in infrared transceiver. Likewise, the HP Omnibook 530 has a 115K baud, bidirectional infrared port. These two devices can link to an HP Vectra 486XM PC, which has a standard infrared port to link to HP mobile products. With the infrared connector, you can seamlessly transfer information you gathered on the road or at a meeting in the office to the desktop, or vice versa.

Specifications

Suggested pricing (U.S. dollars): $1.00 per device
Technology: Infrared
Range: Up to 100 feet
Data rate: 115K baud
Compatibility: Standard 15660 UART processor implementable on an ASIC
        chip

INFRALAN Technologies, Inc.

12 Craig Road
Acton, MA 01720
Voice: Direct dial: 1-508-266-1500
Fax: 1-508-635-0806

Product

INFRALAN

Product overview

INFRALAN offers point-to-point infrared wireless products. It supports a
token-passing network capable of supporting 4- or 16-Mbps (standard
Token Ring speeds) throughput. The system will work with any IEEE 802.5
network operating system and will plug directly into any standard token
network interface card. The system composes a base unit (resembling a
multistation access unit) and two optical nodes. The system devices have
an optical sensor to detect the presence or absence of transmitted light.
Naturally, you need two sensors at each optical node, one for the ring in
(uplink) and one for the ring out (downlink).

Specifications

Technology: Infrared
Range: 80 feet
Frequency: Infrared
Data rate: Up to 16 Mbps
Protocol: Token passing
Compatibility: Banyan VINES, IBM LAN Server, Microsoft LAN Manager,
        Novell NetWare Lite, NetWare 2.x and 3.x, ODI, NDIS,
        TCP/IP and IPX/SPX.

International Business Machines Corporation (IBM)

See your local sales representative.

Product

Infrared Wireless Adapters for Personal Area Network (PAN) System

Product overview

IBM offers a family of infrared adapters that can form a stand-alone wireless LAN. This approach can be used to form personal area networks (PANs), which IBM defines as "LAN environments restricted to physical space."

Cards are available to connect to ISA, MCA, and PCMCIA buses. Each card attaches to a separate infrared transceiver via a short cable. The adapters use diffused infrared light, which bounces off existing walls and ceilings without the use of special reflectors. This limits speed and distance, relative to other infrared techniques, but has the advantage of not requiring special aiming of the transceivers. Limited range also offers improved security of data transfers.

For access to a wired LAN, an infrared adapter could be placed in a PC bridge that also contains a wired LAN card.

Specifications

Technology: Infrared
Range: Coverage up to 30 feet by 30 feet
Frequency: Diffuse infrared
Data rate: Up to 1 Mbps
Protocol: Carrier Sense Multiple Access (CSMA)
Compatibility: ODI, NDIS (planned for 2nd quarter 1994), Novell NetWare
          Lite, NetWare 3.x and 4.x, OS/2 2.1 (planned for 2nd quarter,
          1994)

Laser Communications Inc.

1848 Charter Lane
Suite F
P.O. Box 10066
Lancaster, PA 17605-0066
Voice: Toll free: 1-800-527-3740; Toll: 1-717-394-8634
Fax: 1-717-396-9831

Product

LACE

Product overview

Laser Atmospheric Communication Equipment (LACE) is a family of products that use lasers as a basis for short-haul, optical transmission. Because

LACE uses laser technology, the devices are far superior to infrared transmission devices, especially under high ambient lighting conditions and in poor weather. Using coherent laser light, LACE links are ideally suited for campus settings and urban locations where the installation of cable is impractical or impossible and the performance of leased lines is too slow or costly. You can link communication systems by laser beam up to 1 kilometer apart, provided you have line-of-sight.

LACE can connect Ethernet and Token Ring LANs and replace T1/T2 links. Figure A.8 provides an example of an Ethernet typical application, while Figure A.9 provides an example of LACE as a bridge. Each unit simultaneously transmits and receives signals, so only one system is necessary for full duplex operations. Unlike microwave transmission, LACE does not require government approval.

Specifications

Technology: GaAlAs Diode Laser
Range: Up to 1 kilometer
Frequency: Near-infrared, wave length: 820nm
Data rate: Up to 16 Mbps
Compatibility: Ethernet: IEEE 802.3; Token Ring: IEEE 802.5

Microwave Radio Corporation

20 Alpha Road
Chelmsford, MA 01824-4168
Voice: Domestic: 1-508-250-1110
Direct dial: 1-508-256-8335
Fax: Domestic: 1-508-256-5215
International: 1-508-256-6225

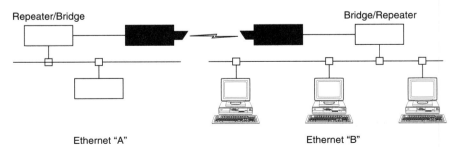

Figure A.8 LACE Ethernet application.

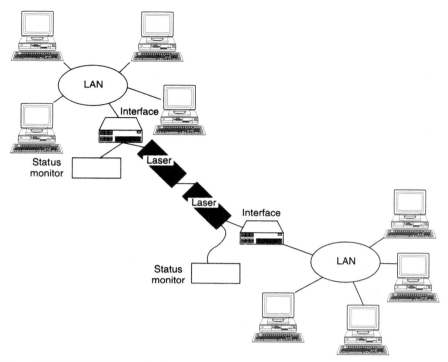

Figure A.9 LACE bridging networks.

Product

Microlink

Product overview

Microwave Radio Corporation (MRC) offers a wide range of data and video transmission products for linking multiple facilities, educational institutions, companies, and other organizations. MRC offers a medium-to-short-haul (up to 10 miles) digital link in the 21.2 to 23.6-GHz range.

The microwave system is ideal for digital connections because it is reliable, easy to install, and affordable. It consists of a lightweight outdoor radio frequency (RF) unit, a precision antenna, and an indoor digital service unit.

Specifications

Technology: Microwave
Range: Up to 10 miles

Frequency: 21.2 to 23.6 GHz
Data rate: Up to 8.448 Mbps

Mitsubishi Electronics America, Inc.

Electronic Device Group
1050 East Arques Avenue
Sunnyvale, CA 94086
Voice: 1-408-730-5900
Fax: 1-408-737-1129

Product

Integrated circuits and wireless components

Product overview

Mitsubishi has RF integration on a single-chip gallium arsenide (GaAs) monolithic microwave integrated circuits (MMICs) for personal communication systems (PCS). The company has long been active in RF applications, with key products in the GaAs MMICs, RF power transmitters, optoelectronic transmitters, receivers, and transceivers. The MMIC is ideal for portable ISM products, including inventory control systems, automatic meter readers, PCMCIA IC cards, and laptop RF modems with wireless LANs.

Specifications

Technology: GaAs MMIC
Frequency: 902–928 MHz (ISM band); 1.85–1.95 GHz

Motorola Wireless Data Group

50 East Commerce Dr.
Suite M1
Schaumburg, IL 60173
Voice: Toll free: 1-800-233-0877
Direct dial: 1-708-576-7164

Product

PCMCIA wireless modem

Product overview

Motorola Inc.'s Wireless Data Group has announced the technology for a new family of intelligent wireless modems for pocket-size, hand-held, and portable computing and communicating devices.

Currently under development, this new family of PCMCIA-format intelligent modems, the first of its kind in the wireless and wireline data market, will enable both one- and two-way, wide- and local-area networking, and it will operate over the most popular public and private networks. The PCMCIA (Personal Computer Memory Card International Association) standard supports credit-card-size peripheral devices that add memory and I/O capabilities to computers.

Motorola wants their wireless modems to work, on a worldwide basis, on a wide range of computing/communicating devices and operating system platforms, and to be compatible with a breadth of software applications. This will enable the development and release of PCMCIA wireless-enabled products such as palmtop, notebook, and pen-based computers, PDAs, and personal communications devices.

Product

Altair VistaPoint and VistaPoint LR

Product overview

Motorola's Altair product line is Ethernet-based, and has a fixed-frequency microwave range using CSMA/CA. The intention of the product line is to provide the highest bandwidth in a controlled radio frequency. Motorola holds the license for the frequency ranges for Altair and authorizes the use of the license in a specific site. Motorola will configure a system operating within a building, warehouse, or plant, and it grants permission to the customer to use the frequencies. To prevent interference from other users, Motorola ensures that the customer is the only user of the frequency within a 5-mile radius.

Motorola's Wireless Data Group recently announced the availability of Altair VistaPoint LR, an extended-distance, high-speed wireless system for reliable inter- and intrabuilding Ethernet transmission. Based on Motorola's existing proven Altair wireless Ethernet LAN technology, they designed Altair VistaPoint LR as a cost-effective alternative to leased telephone lines and conventional microwave technology. It can save money by eliminating monthly charges associated with T1/E1 telephone facilities and save time by eliminating delays involved with licensing and site construction of conventional microwave equipment.

Operating at 5.3 Mbps, Altair VistaPoint LR and standard Altair VistaPoint are viable solutions for LAN applications calling for fractional or full T1/E1 performance to connect buildings. Throughput of the system naturally varies based on packets and protocols. Altair VistaPoint products are most attractive as solutions where buildings are separated by rights of way (such as highways and railroads), natural obstacles (such as rivers), or where there is a need to dig trenches and lay cables.

In addition, Altair VistaPoint products are ideal disaster recovery solutions. They can help keep mission-crucial applications running in the event of a disater, such as a severed cable. With additional hardware, the primary wired facilities can automatically switch over to the Altair VistaPoint system to avoid loss of crucial data after a disaster occurs.

You also can use Altair VistaPoint products to interconnect networks on separate floors of a leased building or for companies that need to temporarily link networks. While Motorola's previously announced Altair Plus II wireless LAN system is an excellent solution for automating the factory or warehouse floor, Altair VistaPoint products are well-suited to linking those environments with the rest of a company's office network. Because the system operates at 18 GHz, the LAN's wireless communications are immune to factory-generated noise.

The Altair VistaPoint LR system consists of two modules: a main module (MM) and a remote module (RM). The MM connects to an Ethernet server or network (wired or wireless), and the RM connects to a remote LAN segment, located either in a different building or a different region within the same building. The system is fully compliant with IEEE Ethernet 802.3 and supports all Ethernet-compatible network operating systems and protocols.

In the United States, standard Altair VistaPoint operates at distances up to 500 feet (150 meters), while Altair VistaPoint LR links buildings up to 3,940 feet (1.2 kilometers). In most other countries, Altair VistaPoint LR can transmit up to 6,900 feet (2.1 kilometers).

Specifications

Suggested pricing (U.S. dollars): Altair VistaPoint: $11,500; Altair VistaPoint LR: $15,000
Technology: Microwave (fixed frequency)
Range: Up to 3940 feet (1.2 kilometers) in the United States
Frequency: 18 GHz
Data rate: 5.3 Mbps
Compatibility: Ethernet: IEEE 802.3
Network management: Simple Network Management Protocol (SNMP) agent support

Norand Data Systems, Ltd.

550 Second Street S. E.
Cedar Rapids, IA 52401
Voice: Toll free: 1-800-452-2757 (product information); 1-800-553-5971 (switchboard)
Direct dial: 1-319-369-3100
Fax: 1-319-369-3453

Product

RT1100 and RT1700 Radio Data Terminals

Product overview

The Radio Data Terminal is for applications requiring large amounts of sales or logistics information displayed. As a result, it is ideal for industrial, warehousing, and retail environments.

The benefit to using a radio data terminal is its modularity. For example, you can purchase a basic UHF terminal and add an integrated laser or CCD scanning module or upgrade to spread-spectrum technology. Interchangeable modules allow you to quickly handle new applications.

Wireless base-to-base communications can be established instantly whenever needed. For speed and data reliability, an automatic configuring capability picks the optimal path for data to travel between the base stations and your host computer.

Specifications

Technology: Spread-spectrum
Range: N/A
Frequency: 902–928 MHz (ISM band, no site licensing required)
Data rate: 192 Kbps
Compatibility: RS-232

O'Neill Connectivity, Inc.

2445 Maryland Road
Willow Grove, PA 19090
Voice: Toll free: 1-800-624-5296
Direct dial: 1-215-957-5408
Fax: 1-215-830-1207

Product

LAWN

Product overview

LAWN uses one transceiver per computer, with separate transceivers for printers and modems. The transceivers can be used as repeaters to extend cabling beyond 100 meters. You can set LAWN up in a cellular architecture where the cells use different spreading codes, but that does not provide the hand-over procedures needed to permit mobile devices to roam throughout the network.

Specifications

Technology: Spread-spectrum
Range: Up to 35 meters
Frequency: 902–928 MHz (ISM band, no site licensing required)
Data rate: 38.4 Kbps
Compatibility: RS-232

Persoft, Inc.

465 Science Drive
P.O. Box 44953
Madison, WI 53744-4953
Voice: Toll free: 1-800-368-5283
Direct dial: 1-608-273-6000
Fax: 1-608-273-8227

Product

Intersect Remote Bridge and Concentrator

Product overview

Since 1991, Persoft has provided spread-spectrum radio frequency wireless products for Ethernet and Token Ring. The Intersect Remote Bridge connectivity solutions provide a high-speed, low-cost alternative. Spread-spectrum provides a data transmission rate of 2 Mbps. Address filtering and access control maintain data security. The Ethernet and Token Ring Bridges filter on protocol, source, and destination addresses. The Intersect Bridge is a MAC layer bridge, providing optimal performance in multivendor environments. It supports the IEEE 802.1D spanning-tree protocol. The Intersect Concentrator gives wireless network users complete Ethernet or Token Ring connectivity, providing access to the wired network. DES encryption is available as an option on the Intersect Remote Bridge and Concentrator.

Specifications

Suggested pricing (U.S. dollars): Ethernet bridge: $12,990 (pair), Token
                                   Ring bridge: $13,990 (pair)
Technology: Spread-spectrum
Range: Up to 800 feet with omnidirectional antenna; up to 3 miles with directional antenna
Frequency: 902–928 MHz (ISM band, no site licensing)
Data rate: 2 Mbps

Compatibility: Protocol independent; Ethernet: IEEE 802.3, Token Ring:
    IEEE 802.5
Network management: Simple Network Management Protocol (SNMP)

Photonics

2940 N. First Street
San Jose, CA 95134
Voice: Direct dial: 1-408-955-7930
Fax: 1-408-955-7950

Product

Photolink

Product overview

Photolink defines a transparent circuit-switched point-to-multipoint network. It enables up to 40 devices to communicate with a central concentrator that compresses the traffic into 16 distinct channels.

The transceivers each connect 4 devices; the concentrator uses a special
high-capacity transceiver. All transceivers must be aimed at a central re-
flecting area such as a ceiling or wall, but not a mirror, because the product
relies on diffuse reflections to distribute the signals. Photolink needs spa-
tially distinct zones, although only the actual reflection surfaces must be
distinct.

Specifications

Technology: Infrared
Range: Up to 10 meters
Data rate: 9.6 Kbps
Compatibility: RS-232

Proxim, Inc.

295 North Bernardo Avenue
Mountain View, CA 94043
Voice: 1-415-960-1630
Fax: 1-415-964-5181

Product

RangeLAN1

Product overview

Proxim Inc.'s RangeLAN1 is the former RangeLAN product, renamed since the introduction of the RangeLAN2 product. The older technology used proprietary frame types and could not support TCP/IP. However, the product came with "roaming" software, which is quite valuable.

The RangeLAN1 product family, operating at 242 Kbps, will continue to be marketed for lower-speed LAN applications such as e-mail, printer sharing, and terminal emulation as well as for mobile computing solutions in vertical markets like health care, manufacturing, distribution, and retailing.

Specifications

Suggested pricing (U.S. dollars): RangeLAN1/ISA: $495; RangeLAN1/LT
(Compaq LTE and Contura notebooks):
$595; RangeLAN1/PCMCIA: $595;
RangeLAN1/Parallel: $595;
Technology: Spread-spectrum (direct sequence)
Range: 300–500 feet indoors, up to 800 feet outdoors
Frequency: 902–928 MHz (ISM band, no site licensing required)
Data rate: 242 Kbps
Protocol: CSMA/CA
Compatibility: LAN Manager, NDIS, NetWare Lite, NetWare 2.x, 3.x and 4.x,
ODI and Windows for Workgroups
Network management: Hardware CRC-16 and Logical Link Control (LLC)

Product

RangeLAN2

Product overview

Proxim, Inc. has announced a new family of long range, high-speed wireless local area networking (LAN) products. RangeLAN2's raw uncompressed data rate is 1.6 Mbps. RangeLAN2 delivers the longest range of any wireless LAN product in its class, operating at a distance of up to 300–500 feet in normal office locations and up to 800–1000 feet in open space environments. In addition, the RangeLAN2 family provides transparent access to standard wired LAN environments.

The RangeLAN2 product line consists of three separate products:

- RangeLAN2/PCMCIA, a high-performance, low-power wireless network adapter for any mobile PC equipped with a Type II Personal Computer Memory Card International Association (PCMCIA) card slot. The PCMCIA card enables portable computer users to access existing wired client/server networks or to communicate on a peer-to-peer basis with other RangeLAN2 desktop and portable systems.

- RangeLAN2/ISA, a wireless LAN adapter card for desktop PCs and servers that fits into the expansion slot of most desktop PCs and servers.
- RangeLAN2/Bridge, a wireless local bridge that provides transparent access to wired LANs.

The RangeLAN2 family ships with drivers that support most standard LAN operating systems, including NetWare, NetWare Lite, LAN Manager, and Windows for Workgroups.

All RangeLAN2 systems come bundled with Proxim's unique RangeLAN2/Roaming software, a breakthrough capability initially developed for Proxim's original RangeLAN family of adapters. RangeLAN2/Roaming allows mobile computer users to move seamlessly from one access point to another without losing connection to the LAN, enabling organizations to create "micro-cellular" in-building wireless networks.

RangeLAN2/PCMCIA will support PCMCIA version 2.0 software, which includes Card and Socket Services for "hot insertion" capability. RangeLAN2/PCMCIA will operate at an average power output of 100 milliwatts, the same power output as the original RangeLAN/PCMCIA, and the lowest power consumption of any wireless LAN product on the market.

RangeLAN2/ISA provides an easy-to-install, wireless extension to existing wired client/server LANs for desktop locations that are difficult or expensive to wire. It also can act as a stand-alone, peer-to-peer networking solution in smaller companies or departments. Any RangeLAN2/ISA device connected to a wired LAN can function as the base station or access point for other RangeLAN2 systems not physically wired to the network.

The RangeLAN2/Bridge is a wireless local bridge that enables transparent access to standard wired LAN environments. Operating at the Media Access Control (MAC) layer of the ISO network model, RangeLAN2/Bridge enables RangeLAN2-equipped portable and desktop computer users to access a wired LAN. RangeLAN2/Bridge currently support Ethernet LANs.

Unlike Proxim's original RangeLAN adapter family, which uses direct sequence spread spectrum technology in the 902–928-MHz frequency band, RangeLAN2 is based on frequency-hopping spread-spectrum technology in the 2.4–2.483-GHz band.

RangeLAN2 delivers a raw uncompressed data rate of 1.6 Mbps per channel with 10 independent channels available. This multichannel, frequency-hopping approach enables 10 independent wireless LANs to operate within the same physical space, effectively increasing the aggregate capacity of RangeLAN2 networks tenfold. Depending on traffic and use, you might add users and subnets in a scalable fashion and load balance across networks. Proxim plans to offer a RangeLAN2 trade-up program for any existing RangeLAN1 user seeking additional wireless LAN performance or functionality.

Specifications

Suggested pricing (U.S. dollars): RangeLAN2/ISA: $595, RangeLAN2/PCMCIA: $695, RangeLAN2/Bridge for Ethernet networks: between $1,500 and $2,500 depending on configuration
Technology: Spread-spectrum (frequency-hopping)
Range: Up to 300–500 feet indoors and up to 800–1000 feet outdoors
Frequency: 2.400–2.483 GHz
Data rate: 1.6 Mbps
Compatibility: ISA, MCA and PCMCIA, ships with drivers that support most standard LAN operating systems, including NetWare, NetWare Lite, LAN Manager, and Windows for Workgroups

Sharp Electronics Corp.

5700 NE Pacific Rim Boulevard
MS 20
Camas, WA 98607-9489
Voice: 1-206-834-8948
Fax: 1-206-834-8903

Product

Serial infrared interface

Product overview

Sharp also has a serial infrared communications interface for communication between personal computer products ranging from hand-held devices to desktop PCs, printers, and other peripheral products such as Wizard organizers and Apple Newton MessagePad.

Specifications

Technology: Infrared
Range: Up to 100 feet
Data rate: 115K baud

Solectek

6370 Nancy Ridge Drive
Suite 109

San Diego, CA 92121-3212
Direct dial: 1-619-450-1220

Product

AirLan/Parallel, AirLan/Hub and AirLan/Bridge

Product overview

AirLan/Parallel is a wireless Ethernet adapter that attaches externally to a parallel port. With a 2 Mbps data rate, AirLan/Parallel is designed for portable PCs or other stand-alones. The device has rechargeable batteries for mobile operations of up to 10 hours without draining power. It can operate up to 800 feet in an open environment and is compatible with the AirLan/Hub.

The AirLan/Hub is an ALR 386/16 with one Ethernet card and one AirLan wireless card. The hub is flexible because it can be used in a number of ways, but it lacks network management software. The AirLan/Bridge also is an ALR 386/16 machine with one Ethernet card, one AirLan wireless card, and bridge software. It is a learning bridge, dynamically learning the machines on either side of the connection.

Specifications

Technology: Spread-spectrum
Range: Up to 800 feet
Frequency: 902–928 MHz (ISM band, no site licensing required)
Data rate: 2 Mbps
Compatibility: RS-232

South Hills Datacomm

760 Beechnut Drive
Pittsburgh, PA 15205
Voice: Toll free: 1-800-245-6125
Direct dial: 1-412-921-9000
Fax: 1-412-921-2254

Product

Wireless Ethernet and Token Ring Bridges

Product overview

These are protocol-transparent bridges creating 2 Mbps wireless links between cabled Ethernet or Token Ring LANs up to 2 miles apart. The Ethernet bridge is IEEE 802.3 compliant, while the Token Ring bridge is IEEE 802.5 compliant. DES encryption can be purchased as an option.

Specifications

Suggested pricing (U.S. dollars): Ethernet: $12,990 (pair); Token Ring:
$13,990 (pair)
Technology: Spread-spectrum
Range: Up to 2 miles
Frequency: 902–928 MHz (ISM band, no site licensing required)
Data rate: 2 Mbps
Compatibility: Protocol independent; Ethernet: IEEE 802.3, Token Ring:
IEEE 802.5
Network management: Simple Network Management Protocol (SNMP)
agent support

Product

Wireless Ethernet Concentrator

Product overview

You can use the wireless Ethernet concentrator to share data between wired and wireless LANs. This concentrator lets you connect any PC with the wireless network cards below to a cabled Ethernet network.

Specifications

Suggested pricing (U.S. dollars): $3,295
Technology: Spread-spectrum
Range: Up to 200 feet indoors with omnidirectional antenna; up to 2 miles
outdoors with directional antenna
Frequency: 902–928 MHz (ISM band, no site licensing required)
Data rate: 2 Mbps
Compatibility: Novell, DECnet, TCP/IP, and XNS

Product

Wireless network cards

Product overview

You can use wireless network cards to provide a wireless interface for a PC or PS/2 to a wireless PC LAN or to a wireless concentrator to connect to a wired LAN. The wireless cards operate at 2 Mbps, are compatible with Ethernet, Token Ring, and ARCnet, and can be used with a DES encryption chip.

Specifications

Suggested pricing (U.S. dollars): $995
Technology: Spread-spectrum
Range: Up to 200 feet
Frequency: 902–928 MHz (ISM band, no site licensing required)
Data rate: 2 Mbps
Compatibility: ISA, MCA, LAN Manager, NDIS, NetWare 2.x and 3.x

Spectrix Corporation

906 University Place
Evanston, IL 60201-3121
Voice: 1-708-491-4534
Fax: 1-708-467-1094

Product

SpectrixLite

Product overview

SpectrixLite provides cost-effective LAN connectivity. The system creates a set of room-based "cells" of coverage using small infrared antenna units mounted in each room. The antenna units are connected to a PC card. A SpectrixLite hand-held communicator entering a cell gains wireless access to the network through the SpectrixLite PC bridging software. As the user moves from room to room, the system transparently maintains conductivity to the resources.

Reliability and bandwidth are inherent to the SpectrixLite product because of the use of infrared. This reliability is enhanced by the use of the CODIAC wireless protocol, which is a deterministic polling approach. Figure A.10 shows a SpectrixLite network compared to standard NetWare Ethernet network.

Specifications

Technology: Diffuse infrared
Range: 50 feet
Frequency: Infrared
Data rate: 4 Mbps
Network management: Simple Network Management Protocol (SNMP)

Wi-LAN Inc.

#308-809 Manning Rd. N.E.
Calgary, AB

T2E 7M9
Voice: Direct dial: 1-403-273-9133
Fax: 1-403-273-5100

Product

Model 902-20 Wireless LAN

Product overview

The Wi-LAN 902-20 Wireless LAN is an external computer device that transmits data at rates up to 20 Mbps from one PC to another using a peer-equal network approach. The 902-20 Wireless LAN replaces the wired link

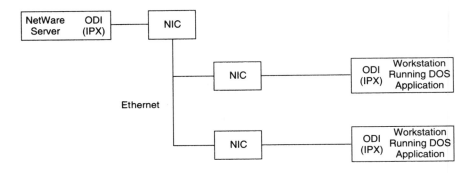

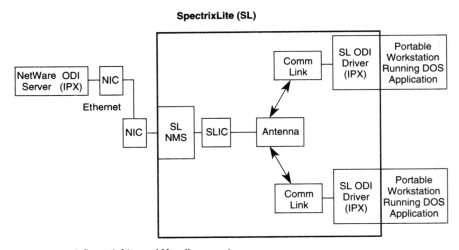

Figure A.10 A SpectrixLite and Novell comparison.

between nodes in local area networks to provide flexibility in movement and position for the nodes.

Specifications

Technology: Spread-spectrum (wideband)
Range: 90 feet in an office environment, 300 feet unobstructed
Frequency: 902–928 MHz (ISM band, no site licensing required)
Data rate: Aggregate network throughput is 20 Mbps, node-to-node
throughput is 1.5 Mbps
Compatibility: IEEE 802.3, 10Base-T, 10Base-5 and AUI

Windata Inc.

10 Bearfoot Road
Northboro, MA 01532-1506
Voice: Direct dial: 1-508-393-3330
Fax: 1-508-393-3694

Product

AirPort I Hub and Remote

Product overview

AirPort I is the first member of Windata's wireless interbuilding products. It can communicate at distances up to 1000 feet with line-of-sight. AirPort I uses Windata's wireless Hub and Remote architecture. The Hub is the master controller and manages the Remote.

Specifications

Suggested pricing (U.S. dollars): Hub: $8,750; Remote: $3,700
Technology: Spread-spectrum
Range: 1,000 feet
Frequency: Transmitter: 5.746–5.830 GHz; Receiver: 2.400–2.484 GHz (ISM
band, no site licensing required)
Data rate: 5.7 Mbps
Compatibility: Ethernet: IEEE 802.3
Network management: Simple Network Management Protocol (SNMP)
agent support

Product

AirPort II Hub and Remote

Product overview

AirPort II is the second member of Windata's wireless interbuilding products. It was designed for applications with longer range requirements of up to 1.448 kilometers or $\%_0$ of a mile. Like AirPort I, AirPort II uses Windata's wireless Hub and Remote architecture. The Hub is the master controller and manages the Remote.

Specifications

Suggested pricing (U.S. dollars): Hub: $13,500; Remote: $8,500
Technology: Spread-spectrum
Range: .9 mile
Frequency: Transmitter: 5.746–5.830 GHz; receiver: 2.400–2.484 GHz (ISM
　　　　　　band, no site licensing required)
Data rate: 5.7 Mbps
Compatibility: Ethernet: IEEE 802.3
Network management: Simple Network Management Protocol (SNMP)
　　　　　　agent support

Product

FreePort

Product overview

There are 3 components in the FreePort Wireless LAN system: the Wireless Hub, the Wireless Transceiver, and SeePort Network Management System. (See Figure A.11.) The FreePort Wireless Hub is the center of the architecture.

Specifications

Suggested pricing (U.S. dollars): Hub: $7,450; Remote: $1,030
Technology: Spread-spectrum
Range: 80 meters
Frequency: Transmitter: 5.746–5.830 GHz; Receiver: 2.400–2.484 GHz (ISM
　　　　　　band, no site licensing required)
Data rate: Rated at 12 Mbps, throughput 5.7 Mbps
Compatibility: Ethernet: IEEE 802.3
Network management: Simple Network Management Protocol (SNMP) agent
　　　　　　support, SeePort Network Management System

Figure A.11   FreePort family.

Xircom, Inc.

26025 Mureau Road
Calabasas, CA 91302
Voice: Direct dial: 1-818-878-7600
Sales support: 1-800-438-4526
Fax: 1-818-878-7630
FactsLine: 1-800-775-0400

Product

Netwave Cordless LAN products: CreditCard Netwave Adapter, Pocket
Netwave Adapter, Netwave Access Point for Ethernet. (See Figure A.12.)

Product overview

The Netwave adapters can be used to form a stand-alone NetWave cordless
LAN or gain access to an existing LAN through a Netwave access point.
They use frequency-hopping spread-spectrum technology in the 2.4-GHz
band to achieve a claimed uncompressed data transfer rate of 1 Mbps.
Xircom specifies their range at up to 150 feet indoors, or up to 750 feet in
open space.

The CreditCard adapter is compliant with PCMCIA Release 2.1 and will
fit in a Type II slot. Xircom's Pocket adapter attaches to the computer's parallel port and does not require a separate ac power supply. Each includes an
integrated antenna.

The adapters can be used in combination to form a stand-alone LAN. However, with the addition of a Netwave access point, they can also be linked to an existing wired Ethernet LAN through 10Base-T (unshielded twisted pair) or 10Base-2 (thin coax). Each access point can support 10 to 15 cordless users simultanteously. And with multiple Netwave access points, an adapter user can roam freely throughout the total coverage area. At the heart of the access point is a CreditCard Netwave adapter, which eases future upgrades.

Specifications

Technology: Frequency-hopping spread-spectrum
Range: Up to 150 feet indoors
Frequency: 2.4000–2.4835 GHz
Data rate: Rated at 1 Mbps (uncompressed)
Compatibility: NDIS, ODI, Artisoft LANtastic (NDIS), Banyan VINES, Microsoft LAN Manager, Microsoft Windows for Workgroups, Novell NetWare, Novell Personal NetWare, TCP/IP.

There sure are a slew of wireless vendors in the local area networking marketplace. Evaluating all the vendors could easily become a daunting task. Perhaps understanding what other organizations are buying might help you narrow your search. Figure A.13 shows the results of a Business Research Group 1993 study—offered without prejudice. Several vendors in the list only started to offer products in the last 12 months, so use the chart with care. As you learned in chapter 11, you must understand your organization, evaluate its needs, and select the product that helps you meet those needs.

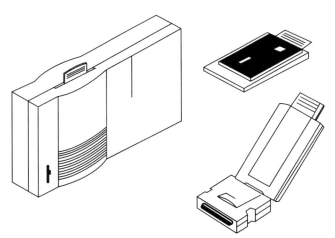

Figure A.12 Xircom's Netwave access point for Ethernet, PCMCIA adapter, and pocket adapter.

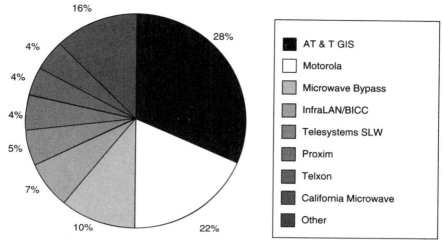

Figure A.13   Wireless LAN market share, 1993.

Wireless Vendors

TABLE B.1 Wireless Cards for PCs

Company	Product	Adapter type	Speed	Frequency	Drivers
AIRONET Wireless Communications, Inc. 3330 West Market Street P.O. Box 5292 Akron, OH 44334-0292 800-800-8016	ARLAN 655, 670 and 690 Wireless Network Interface Cards	655: ISA, 670: MCA, 680: Parallel, 690: PCMCIA	1.35 Mbps	902–928 MHz	Protocol independent
ALPS 3553 North First St. San Jose, CA 95134 408-432-6000	RadioPort Plus	Parallel Port	242 Kbps	902–928 MHz	LANtastic/AI NetWare Lite NetWare Windows for Workgroups
AT&T Global Information Systems 1700 South Patterson Blvd. Dayton, OH 45479	WaveLAN	ISA, MCA, PCMCIA	2 Mbps	902–928 MHz and 2.4 GHz	ISA, LAN Manager, NDIS, NetWare Lite, NetWare 2.x, 3.x and 4.x, ODI, TCP/IP and Windows for Workgroups
Black Box Corp. P.O. Box 12800 Pittsburgh, PA 15241 800-355-8001	BestLAN2	ISA,MCA, PCMCIA	2 Mbps	902–928 MHz	NetWare ODI NDIS
Communications Research & Development Corporation 7210 Georgetown Rd. Suite 300/400 Indianapolis, IN 46268 317-290-9107	Wireless Data Link	RS-232 serial port	38,400 baud	Host: 905.050–908.950 MHz, remote: 923.050–926.950 MHz	Protocol independent

Vendor	Product	Interface	Speed	Frequency	Works with
Granite Communications 9 Townsend West Suite 1 Nashau, NH 03063 603-881-8666	HI5	Parallel port	Up to 121 Kbps	902–928 MHz (ISM band, no site licensing)	VIDEOPAD Links
Proxim, Inc. 295 North Bernardo Ave. Mountainview, CA 94043 800-229-1630	RangeLAN1/RangeLAN2	ISA, Parallel Port, PCMCIA	242 Kbps/1.6 Mbps	902–928 MHz/2.400–2.483 GHz	NetWare Lite NetWare ODI NDIS
Solectek 6370 Nancy Ridge Dr. Suite 109 San Diego, CA 92121 619-450-1220	AIRLAN	ISA, PCMCIA	2 Mbps	902–928 MHz	NetWare NetWare ODI NDIS LANtastic/AI Windows for Workgroups
South Hills Datacomm 760 Beechnut Dr. Pittsburgh, PA 15205 412-921-9000	Wireless Network Cards	ISA, MCA	2 Mbps	902–928 MHz	NetWare NetWare ODI NDIS OS/2
Spectrix Corp. 906 University Place Evanston, IL 60201 708-491-4534	SpectrixLite Interface Card	ISA, PCMCIA	4 Mbps	Diffuse Infrared	NetWare NetWare ODI PDS (TCP/IP)
Telesystems 3330 West Market St. P.O. Box 5582 Akron, OH 44334-0582 800-800-8008	ARLAN 650 & 670	ISA, MCA	1 Mbps	902–928 MHz	NetWare PDS (TCP/IP)

TABLE B.1 Continued.

Company	Product	Adapter type	Speed	Frequency	Drivers
Telesystems 3330 West Market St. P.O. Box 5582 Akron, OH 44334-0582 800-800-8008	ARLAN 680	Parallel Port	1 Mbps	902–928 MHz	NetWare PDS (TCP/IP)
Windata Inc. 10 Bearfoot Road Northboro, MA 01532-1506 508-393-3330	FreePort Transceiver	Parallel Port	5.7 Mbps	2.400–2.484 GHz 5.746–5.830 GHz	

TABLE B.2 Wireless Hubs and Bridges

Company	Product	Speed	Frequency	Best application
AIRONET Wireless Communications, Inc. 3330 West Market Street P.O. Box 5292 Akron, OH 44334-0292 800-800-8016	ARLAN 610 and 610E Wireless Ethernet Hub and ARLAN 620 and 620E Wireless Ethernet Bridge	1.35 Mbps	902–928 Mbps	Hub for remote workgroups or for areas where running wires is difficult or costly. Bridge for connecting two Ethernet cable LANs using spread-spectrum technology.
AT&T Global Information Systems	WavePoint	2 Mbps	902–928 MHz and 2.4 GHz	Access point, similar to hub.
Black Box Corp. P.O. Box 12800 Pittsburgh, PA 15241 800-355-8001	BestLAN2 Hub/Bridge	2 Mbps	902–928 Mbps	Hub for remote workgroups or for areas where running wires is difficult or costly. Bridge for connecting two Ethernet cable LANs using spread-spectrum technology.
Granite Communications 9 Townsend West Suite 1 Nashau, NH 03063 603-881-8666	BS5	Up to 121 Kbps	902–928 MHz	Base station for VIDEOPAD Links system, good for applications with low-speed transmissions, but high activity.
INFRALAN Technologies, Inc.	INFRALAN	Up to 16 Mbps	Infrared	Best as a point-to-point bridge for Token Ring network.

TABLE B.2 Continued.

Company	Product	Speed	Frequency	Best application
Motorola 50 East Commerce Dr. Suite M1 Schaumburg, IL 60173 800-233-0877	Altair Plus II	5.7 Mbps	18–19 GHz	Good alternative to link remote areas to the network. Performance is very good.
Motorola 50 East Commerce Dr. Suite M1 Schaumburg, IL 60173 800-233-0877	Altair VistaPoint	5.3 Mbps	18–19 GHz	Good alternative to T1 link. Performance is good compared with other products that have similar functionality.
O'Neill Communications	LAWN	38.4 Kbps	902–928 MHz	Best as a repeater, or for peripheral sharing.
Persoft, Inc. 465 Science Dr. P.O. Box 44953 Madison, WI 53744-4953 608-273-6000	Intersect Remote Bridge/ Concentrator	2 Mbps	902–928 MHz	For remote workgroups or for areas where running wires is difficult or costly. DES encryption available.
Photonics	Photolink	9.6 Kbps	Infrared	Best for circuit-switched point-to-multipoint network, that is, as a concentrator.
Solectek 6370 Nancy Ridge Dr. Suite 109 San Diego, CA 92121 619-450-1220	Wireless Bridge	2 Mbps	902–928 MHz	Good alternative to a T1 link for networking buildings together.

Vendor	Product	Speed	Frequency	Description
Solectek 6370 Nancy Ridge Dr. Suite 109 San Diego, CA 92121 619-450-1220	Wireless Hub	2 Mbps	902–928 MHz	For remote workgroups or for areas where running wires is difficult or costly.
South Hills Datacomm 760 Beechnut Dr. Pittsburgh, PA 15205 412-921-9000	Wireless Bridges	2 Mbps	902–928 MHz	Connect Ethernet and Token Ring cable LANs and remote devices using spread-spectrum technology.
South Hills Datacomm 760 Beechnut Dr. Pittsburgh, PA 15205 412-921-9000	Wireless Ethernet Concentrator	2 Mbps	902–928 MHz	Provides the link between wireless and wired networks, allowing complete sharing of data and resources.
Telesystems 3330 West Market Street P.O. Box 5582 Akron, OH 44334-0582 800-800-8008	ARLAN 610	1 Mbps	902–928 MHz	Connect Ethernet cable LAN and remote devices using spread-spectrum technology.
Telesystems 3330 West Market Street P.O. Box 5582 Akron, OH 44334-0582 800-800-8008	ARLAN 620	1 Mbps	902–928 MHz	Connect two Ethernet cable LANs using spread-spectrum technology.
Windata Inc. 10 Bearfoot Road Northboro, MA 01532-1506 508-393-3330	FreePort Hub	5.7 Mbps	2.400–2.484 GHz 5.746-5.830 GHz	Connect two Ethernet cable LANs.

TABLE B.3 Wireless Links

Company	Product	Speed	Frequency	Best application
Cylink 310 N. Mary Ave. Sunnyvale, CA 94086 408-735-5800	AirLink	256 Kbps	902–928 MHz	Good for point-to-point wireless connectivity.
Cylink 310 N. Mary Ave. Sunnyvale, CA 94086 408-735-5800	AirLink Multipoint	64 Kbps	902–928 MHz	Good for point-to-multipoint wireless connectivity.
Laser Communications Inc. 1848 Charter Lane Suite F Lancaster, PA 17605-0066 717-394-8634	LACE Omnibeam 2000	16 Mbps	Near-infrared	Link communication systems by laser beam to connect sites up to .6 miles (1 km). Good alternative to a T1 link for networking buildings together.
Microwave Radio Corporation 20 Alpha Road Chelmsford, MA 01824-4168 508-250-1110	MR-22DR-NS	8.448 Mbps	21.2–23.6 GHz	Good for wireless digital connections, such as campus area networks up to 10 miles (16 km).
Wi-LAN Inc. #308-809 Manning Rd., N.E. Calgary, AB T2E 7M9 403-273-9133	Model 902-20 Wireless LAN	20 Mbps	902–928 MHz (ISM band, no site licensing required)	Good replacement for the wired link between nodes in a LAN.
Windata Inc. 10 Bearfoot Road Northboro, MA 01532-1506 508-393-3330	AirPort I & II	5.7 Mbps	2.400–2.484 GHz 5.746–5.830 GHz	Good for wireless interbuilding connections, such as campus area networks up to .9 miles (1.4 km).

Frequency Bands

Table C.1 provides the frequencies assigned from the electromagnetic spectrum. The ones used for wireless LAN communication are 902.000–928.000, 2400.000–2483.500, and 5750.000–5825.000 MHz. These are presently available for the "unlicensed operation of approved radio devices." Formerly these bands were classified as industrial, scientific, and medical (ISM).

This spectrum also shows you AM radio (.535–1.705 MHz), FM radio (88.000–108.000 MHz), TV Channels 2–13 (54.000–88.000 and 174.000–216.000 MHz), TV Channels 14–69 (470.000–806.000 MHz), cellular phone (824.000–849.000 and 870.000–894.000 MHz), and police radar (10525.000, 24150.000–34200.000, and 35200.000 MHz).

TABLE C.1 Assigned Frequencies

From . . .	To . . .	Description
0.535	1.705	AM Broadcast Standard North America AM
1.705	30.000	Shortwave, amateur, CB, Ship-to-shore
30.010	37.000	Government, Army/Navy/Coast Guard, Commercial
37.020	37.420	Police mobile, Police, Public Service
37.440	39.000	Government
39.020	39.980	Police, Public service, Police mobile
40.010	41.990	Government, Army/Navy/Air Force
42.020	42.940	Police/Police mobile
42.960	44.600	Commercial
44.620	46.580	Fire, Forestry Conservation, Highway Maintenance, Police, Public Service

TABLE C.1 Continued.

From . . .	To . . .	Description
46.610	47.000	Air Force/Army, Government
47.020	47.400	Highway Maintenance
47.420		Red Cross
47.440	49.580	Special Industry, Special Emergency, Power and Water, Forest Products/Petroleum Products
49.610	49.990	Air Force/Army, Government
50.000	54.000	Amateur
54.000	72.000	VHF TV Channels 2-4
72.000	73.000	Intersystem, Paging, RC, Astronomy
76.000	88.000	VHF TV Channels 5-6
88.000	108.000	FM Broadcast
108.020	132.000	Commercial aircraft
136.000	138.000	Satellite
150.775	160.200	Ambulance, Fire, Police, Forest Conservation, Taxi Base/Mobile, Commercial/Industrial, Utilities, Local Government Mobile, Auto/Special Emergency, Highway Maintenance Mobile, Maritime
160.215	161.565	Railroad
174.000	216.000	VHF TV Channels 7-13
225.000	381.000	Military aircraft
406.000	408.000	Digital data transmission
420.000	470.000	Amateur, Broadcast Pickup, Water and Power, Commercial/Industrial, Telephone Maintenance, Taxi, Motor Carrier/Railroad, Auto Club, Press/Newspapers, Local Government, Public Safety, Commercial, Medical, Police
470.000	806.000	UHF TV Channels 14-69
824.000	349.000	Cellular Mobile Telephone—Mobile
850.000	870.000	Trunk radio systems
870.000	894.000	Cellular Mobile Telephone—Cell site
902.000	928.000	Unlicensed operation of approved radio devices
928.0125	932.000	Domestic Public Radio Service
952.100	959.800	Private Microwave Service
959.8625	959.9875	Common Carrier Radio Service Wide Area Paging
1240.000	1300.000	Amateur
2400.000	2483.500	Unlicensed operation of approved radio devices
3300.000	3500.000	Amateur
5750.000	5825.000	Unlicensed operation of approved radio devices
5925.000	6875.000	Common Carrier and Fixed SAT
8800.000	8800.000	Airborne Doppler Radar
10000.000	10500.000	Amateur
10525.000	10525.000	Police X-band radar gun
24000.000	24250.000	Amateur
24150.000	24150.000	Police k-band radar gun
34200.000	35200.000	Police Ka-band photo-radar gun
48000.000	50000.000	Amateur
71000.000	76000.000	Amateur
165000.000	170000.000	Amateur
240000.000	250000.000	Amateur
300000.000	and above	Amateur

D

Organizations

Following are some organizations who are directly or indirectly concerned with wireless data communications or local area networking.

American Society for Industrial Security (ASIS)
1655 N. Fort Meyer Drive, Suite 1200
Arlington, VA 22209

Association for Computing Machinery (ACM)
11 W. 42 St.
New York, NY 10036

Association for Information Management (AIM)
7380 Parkway Drive
La Mesa, CA 92401

Association for Systems Management (ASM)
P.O. Box 38370
Cleveland, OH 44138-0370

Association of Contingency Planners (ACP)
P.O. Box 73-149
Long Beach, CA 90801-0073

Bank Administration Institute (BAI)
60 Gould Center
Rolling Meadows, IL 60008

Business Resumption Planners Association (BRPA)
P.O. Box 1078
Niles, IL 60648-5078

Business and Industry Council for Emergency Planning and Preparedness (BICEPP)
P.O. Box 9457
Newport Beach, CA 92658

Canadian Information Processing Society (CIPS)
430 King Street West, Suite 205
Toronto, ON
M5V 1L5

Canadian Standards Association (CSA)
178 Rexdale Boulevard
Rexdale, ON
M9W 1R3

CDPD Industry Input Coordinator
650 Town Center Drive
Suite 820
Costa Mesa, CA 92626

Cellular Telecommunications Industry Association (CTIA)
1133 21st Street, NW, Third Floor
Washington, DC 20036

Citizens Concerned About EMFs
P.O. Box 120
San Ramon, CA 94583

COMMON
401 North Michigan Avenue
Chicago, IL 60611

Computer and Business Equipment Manufacturers Association (CBEMA)
311 First St. NW, Suite 500
Washington, DC 20001

Computer Professionals for Social Responsibility (CPSR)
P.O. Box 717
Palo Alto, CA 94301

Computer Security Institute (CSI)
360 Church St.
Northboro, MA 01532

CDPD Industry Input Coordinator
650 Town Center Dr., Suite 820
Costa Mesa, CA 92626

Data Administration Management Association (DAMA)
152 W. Northwest Highway, Suite 103
Palatine, IL 60067

Data Entry Management Association (DEMA)
101 Merrit, 7 Corporate Park
Norwalk, CT 06851

Data Processing Management Association (DPMA)
505 Busse Highway
Park Ridge, IL 60068-3191

Digital Equipment Computer Users Society (DECUS)
219 Boston Post Road (BP02)
Marlboro, MA 01752

Disaster Recovery Institute
5647 Telegraph Road
St. Louis, MO 630129

EDP Auditors Association (EDPAA)
P.O. Box 88180, 300 Schmale
Carol Stream, IL 60188-0180

Electromagnetic Pollution Society (EMPAS)
c/o Frank and Darlene Kafka
P.O. Box 60
Burns Lake, BC
V0J 1E0

Environmental Protection Agency (EPA) EMF Group
Office of Radiation Programs
401 M Street, SW
Washington, DC 20460

European Telecommunications Standards Institute
F-06921 Sophia Antipolis Cedex
Route des Lucioles
France

GUIDE
401 North Michigan Avenue
Chicago, IL 60611

IEEE
(*See Appendix E*)

Information System Security Association (ISSA)
401 North Michigan Avenue
Chicago, IL 60611

Institute of Internal Auditors (IIA)
249 Maitland Avenue, Box 1119
Altamonte Springs, FL 32701

National Computer Security Association (NCSA)
227 West Main Street
Mechanicsburg, PA 17055

Office of Technology Assessment
Wireless Project Team
Telecommunications and Computing Technologies Program
U.S. Congress
Washington, DC

SHARE
401 North Michigan Avenue
Chicago, IL 60611

Where to Get
Information Updates

As you have probably gathered by this point in the book, the world of wireless LANs is constantly evolving. By the time you read this, there will no doubt be new technologies, enhanced product offerings, and changes and enhancements to the standards. Keeping up-to-date on the latest developments is a difficult task.

Fortunately, technology also offers some assistance. If you have access to the worldwide Internet, the latest information on wireless standards and products can be as close as an electronic mail message or newsgroup posting. Selected details from the IEEE working standards group, as well as a general discussion on wireless standards, are both available from the Internet. To give you a taste of what's available, and to help you obtain updates when you need it, we've included the following reference information.

IEEE P802.11 Working Standards Group

Mr. Victor Hayes (Internet e-mail address: Vic.Hayes@Utrecht.NCR.COM), Chairman of the IEEE P802.11 Working Standards Group, and a member of AT&T Global Information Solutions in The Netherlands, very kindly sent us, via electronic mail, the following details on how to obtain information on the latest work of the committee:

> "All documents submitted to and produced by standards working group 802.11 are available to the public subject to restriction by copyright laws.

The working group may decide that certain submissions are not fit for distribution, for instance documents that were only of importance to (a subset of) the attendees of a meeting. All other documents are distributed to the membership of 802.11.

For the general public, all documents distributed to members, but not protected by copyright law, can be ordered from the IEEE 802 document order service. In addition, the public can subscribe to a document subscription service to obtain the documents automatically and nearly as fast as members would receive them. You can contact both services for the latest information through telephone number +1 602 863 0999 or fax +1 602 866 8801.

Many of the documents are also available electronically at anonymous ftp servers available on the Internet. With the document server, we have a fast way of making documents available. However, the authoritative versions remain the paper versions.

There are two anonymous ftp sites for documentation:

1. Server: atg.apple.com (17.255.4.30), maintained by Kerry Lynn of Apple (kerlyn@apple.com).

2. Server: ftp.cs.utwente.nl (130.89.10.247), provided by Ronald Brockman, University of Twente (roana@cs.utwente.nl).

   See the directory

   ```
   /pub/802.11
   ```

   for the

   ```
   readme
   ```

   file. Also see the directory

   ```
   /pub/802.11/ibmpc
   ```

   to find the subdirectories

   ```
   1991_docs, 1992_docs, 1993_docs, 1994_docs, bulk_archive, etc,
   issues, std
   ```

   and others.
   The directory

   ```
   etc
   ```

   contains the venue and agenda for future meetings, while the directory

   ```
   issues
   ```

contains the issues document. The (working) draft standard is available on the server but has been encrypted due to copyright constraints. Only voting members will receive the password via the e-mail reflector."

Internet newsgroup

In addition to the ftp servers and contact numbers noted above, the Internet newsgroup

```
comp.std.wireless
```

provides a forum for questions and discussion of the latest on wireless LANs. As is the case for many subjects discussed on the Internet, this moderated newsgroup includes a Frequently Asked Questions (FAQ) file. The FAQ, which will be updated periodically, provides details on the purpose of the newsgroup, as well as additional sources of information. The latest version of the FAQ available at the time of writing is reproduced in the following.

```
Frequently Asked Questions (FAQ) for the newsgroup 'comp.std.wireless'

        This is the very first edition.  Additions and corrections
        are requested.

        Version: 1.0
        Last-modified: June 6, 1994
        Compiled by W. Stuart Jones
        Address comments to std-mod@wes.mot.com (W. Stuart Jones)

        Copyright (C) 1994 by Wesley Stuart Jones.  All rights
        reserved.  Redistribution of this file in both electronic and
        printed form, is permitted provided if this file is distributed
        in its entirety, including this copyright notice.  If you
        redistribute this file, please notify W. Stuart Jones so that
        distribution records can be kept.

Disclaimer:  Any opinions implied or stated in the following document
             are those solely of Wesley Stuart Jones and not his
             employer Motorola, Inc.

                    Table of Contents
------------------------------------------------------------------

    I.  General Information
        1.  What is the purpose for the newsgroup?
        2.  Who is the Moderator?
        3.  What is the purpose of this FAQ?
        4.  What are the rules for posting?
        5.  How do I post to the newsgroup?
        6.  Where is the Archive for the newsgroup?
        7.  What does this stand for?  (List of common abbreviations)

    II. Committee Information
        1.  What are the current committees discussing wireless data
            standards and how can I contact them?
```

III. More Information
   1. Who are the current suppliers of wireless data products and how do I contact them?
   2. How can I get additional information about wireless products/services/companies?
   3. What are good magazine references for wireless data information?

IV. Acknowledgments

```
===========================
```
   I. GENERAL INFORMATION
```
===========================
```

1.1 WHAT IS THE PURPOSE FOR THE NEWSGROUP?

The primary goal of this newsgroup is to promote consistency and mutual awareness between existing committees (e.g., IEEE 802.11 and ETSI RES-10) and any future committees. The newsgroup expedites their deliberations by providing a convenient forum for technical discussion between their scheduled meetings. Additionally, the open nature of the newsgroup helps the standardization process by allowing people not normally connected with the standards groups to provide their input. The newsgroup has no official standing with any standards group or other organization.

A secondary goal is to explore applications for wireless network technology. The emerging standards must, after all, correctly anticipate these applications.

1.2 WHO IS THE MODERATOR?

The moderator for the newsgroup is Wesley Stuart Jones. Mr. Jones is an employee of Motorola, Inc. in the Wireless Data Group. The moderator can be reached by sending electronic mail (email) to 'std-mod@wes.mot.com'. Motorola desires that the newsgroup be operated in a neutral, non-commercial manner which is fair to all.

1.3 WHAT IS THE PURPOSE OF THIS FAQ?

This document, the FAQ for the newsgroup, was composed to provide general information about the newsgroup. It would be impossible to attempt to answer all common questions regarding the field of wireless data communication. Hopefully, the information contained in this document will direct the reader to adequate sources of information answering questions related to the field. Should some questions repeatedly appear in the newsgroup in the future they will be added to this document.

1.4 WHAT ARE THE RULES FOR POSTINGS?

Generally, anything related to any of the standards committees or wireless technologies is welcome. The moderator may reject postings he deems unprofessional or tasteless.

1.5 HOW DO I POST TO THE NEWSGROUP?

Posting to the newsgroup is the same as any other USENET newsgroup. All postings are forwarded to the moderator for approval. If the message does not show up in a day or so and no rejection notice is received, please try again or email a copy of the post to 'std-mod@wes.mot.com'. Those who do not have access to posting facilities please email messages to 'std-mod@wes.mot.com'.

1.6 WHERE IS THE ARCHIVE FOR THE NEWSGROUP?

To the moderator's knowledge there exists no archive for the newsgroup. Anyone who has the ability and is willing to archive the newsgroup please email the moderator.

1.7 WHAT DOES THIS STAND FOR? (A LIST OF COMMON ABBREVIATIONS)

Abbreviations:

ARQ	Automatic Repeat Request
bbs	bulletin board system
bps	bits per second
CCITT	International Telegraph and Telephone Consultative Committee
CDPD	Cellular Digital Packet Data
CSMA	Carrier Sense Multiple Access
comm	communication(s)
DSSS	Direct Sequence Spread Spectrum
EU	European Union
FHSS	Frequency Hopping Spread Spectrum
ISDN	Integrated Services Digital Network
IR	InfraRed
LOS	Line Of Sight
MAC	Medium Access Control
mgmt	management
mgr	manager
NIC	Network Interface Card
NTIA	National Telecommunications & Information Administration
PDA	Personal Digital Assistant
PHY	Physical layer
QoS	Quality of Service
rcv	receive
RF	radio frequency
SS	Spread Spectrum
s/w	software
WL	wireless

```
============================
 II.  Committee Information
============================
```

2.1 WHAT ARE THE CURRENT COMMITTEES DISCUSSING WIRELESS DATA STANDARDS AND HOW CAN I CONTACT THEM?

IEEE P802.11
Working to develop a MAC and PHY specification for wireless connectivity for fixed, portable, and moving stations within a local area with raw bit rates exceeding 1 Mbit/s.
Chairman: Vic Hayes (vic.hayes@utrecht.ncr.com)

ETSI RES-10
The EU has allocated hundreds of MHz in the 5 to 17 GHz bands for
Hiperlan, the wireless data system standard being written by ETSI
RES-10. Time-bounded service will be an integral (non-optional)
part of the standard. User data rates should be 20 Mbps or higher.
Participation in RES-10 is limited to ETSI-members --- the main
requirement is an European address.
Chairman: Bernard Bourin (bourin@dassault-at.fv)

```
===========================
```
III. More Information
```
===========================
```

3.1 WHO ARE THE CURRENT SUPPLIERS OF WIRELESS DATA PRODUCTS AND HOW
     DO I CONTACT THEM?

     The best source for wireless suppliers is Anthony Stieber
     (anthony@alpha1.csd.uwm.edu). Mr. Stieber keeps lists of wireless
     suppliers at the site 'csd4.csd.uwm.edu' in the
     /pub/portables/wireless directory for "anonymous" ftp. The lists
     divided into CDPD, cellular, IR, LAN, paging, satellite, WAN, and
     miscellaneous categories.

3.2 HOW CAN I GET ADDITIONAL INFORMATION ABOUT WIRELESS PRODUCTS/
     SERVICES/COMPANIES?

     The following sites on the Internet have documents which can be
     "ftp-ed". If the login is not stated use "anonymous" to get
     access to the site.

     Format:
     <Name of Source>- <Domain Name (IP Address)>
     <Owner or SysOp>
     <Directory> login: <login> *only if login is not anonymous
     <Description>

     802.11 Documents- atg.apple.com (17.255.4.30)
     Kerry Lynn (kerlyn@apple.com)
     /pub/802.11   (README file and other subdirectories)
     /pub/doc/802.11/ibmpc   (may subdirectories)

     802.11 Documents- ftp.cs.utwente.nl (130.89.10.247)
     (mirror of Apple 802.11)
     Ronald Brockman (roana@cs.utwente.nl)
     /pub/doc/802.11   (README file and other subdirectories)
     /pub/doc/802.11/ibmpc   (many subdirectories)
     IEEE P802.11 (Wireless LAN) documents

     FCC- ftp.fcc.gov (192.104.54.2)
     /pub
     FCC UPCS rules

     University of Aachen- ftp.dfv.rwth-aachen.de (137.226.4.111)
     /pub/doc/cellular/cn login: ftp
     GSM Docs

     NTIA- ntiabbs.ntia.doc.gov (??.)
     N/A - bbs format
     Government information concerning frequency use and allocation

     ??? cs.dal.ca ()
     /comp.archives/comp.dcom.telecom

misc. archives and Qualcomm specs
UWM Info-Computing Services Division- csd4.csd.uwm.edu (129.89.7.4)
Anthony Stieber (anthony@alpha1.csd.uwm.edu)
/pub/Portables/wireless
Lists of contacts, specifications, and descriptions of wireless
networking systems.

??? ftp.comp.lancs.ac.uk (148.88.16.9)
/pub/mpg
mobile computing bibliography

TCP/IP mobility- parcftp.xerox.com ()
/pub/mobile-ip
archives from mobile-IP mailing list

The following sources have documents available if the sources
are commercial.

CDPD Industry Input Coordinator
        Pittaglio Rabin Todd & McGrath
        650 Town Center Dr., Suite 820
        Costa Mesa, CA 92626
        714-545-9400 X235

CDPD specification
        McCaw Cellular
        Clarissa Jurgenson
        Kirkland, WA
        800-42-MCCAW 206-827-4500

Cellular Digital Packet Data system specification
        CDPD Industry Input Coordinator
        650 Town Center Drive, Suite 820
        Costa Mesa, CA 92626

National Telecommunications & Information Administration (NTIA) BBS
        202-482-1199 (bbs number)
        SysOp: Roger Clark
        202-482-1407 (voice)
        email: rclark@ntia.doc.gov

Overnet BBS - News, information, FCC/NTIA reports, discussion
areas, etc.
        P.O. Box 8551
        Calabasas, CA 91372-8551
        SysOp: Scott Goldman (Goldman@overnet.com) For information
        send request to info@overnet.com

If you know of other good sources of information please email them
to the moderator.

3.3   WHAT ARE SOME GOOD MAGAZINE REFERENCES FOR WIRELESS DATA
      INFORMATION?

Communications Week
        PO Box 1094
        Skokie, IL 60076
        708-647-6834 708-647-6838 fax

Data Communications Magazine
        McGraw-Hill Inc.
        1221 Avenue of the Americas
        New York, NY 10020

    EDN
        275 Washington Street
        Newton, MA 02158

    LAN Times
        1900 O'Farrell Street, Suite 200
        San Mateo, CA 94403
        415-513-6800   415-513-6819 fax

    Wireless Design and Development
        Gordon Publications, Inc. (division of Cahners)
        P.O. Box 650
        Morris Plains, NJ 07950
        201-292-5100   201-898-9281 fax

===========================
IV.   ACKNOWLEDGMENTS
===========================

    John McKown - Invaluable feedback and contributions for this document
            and constant advice for the moderator of the newgroup.

    Anthony Stieber - For contributing lots of information to this
            document.

Glossary

absorption The opposite of radiation. The soaking up of light, heat, sound, and other forms of energy by any substance.

access The ability and the means necessary to approach, to store, or to retrieve data. To communicate with and to make use of any resource of a computer system.

accuracy Having no errors. Correct. Exact. Faithful. Precise. Proper. Right. True. Veracious. Vericidal.

ad hoc network A temporary, stand-alone wireless network without a base station controller. Formed by a small group, such as during a meeting.

algorithm A step-by-step procedure, usually mathematical, for doing a specific function, e.g., a PIN verification algorithm or an encryption algorithm.

alternating current (ac) An electrical current that changes strength and direction of flow with a certain regular cycle.

ammeter Device for measuring electric currents.

ampere (amp) A unit used to measure electrical current.

amplitude The height of any wave, whether the wave is sound, water, or electromagnetic.

AMPS (Advanced Cellular Mobile Services) A widely accepted CDPD technology specified by the Electronic Industries Association's TIA IS-9B.

analog A system based on a continuous ratio, such as voltage or current values.

analog transmission A communications scheme using a continuous signal, varied by amplification. Broadband networks use analog transmissions.

antenna A transceiver.

application The user's communication with the installation. A software program or program package allowing a user to perform a specific job, such as word processing or electronic mail.

application program/software A program written for or by a user that applies to his or her work.

application system A collection of programs and documentation used for an application.

architecture The general design of hardware or software, including how they fit together.

ARCnet (Attached Resource Computer Network) A local area network scheme developed by Datapoint.

ARQ Automatic repeat request.

asynchronous A method of data communication where transmissions are not synchronized with a signal. Local area networks transmit asynchronously.

ANSI Acronym for American National Standards Institute, which sets standards for many technical fields.

ASCII Acronym for American Standard Code for Information Interchange. Pronounced "ASK-ee".

ATM Acronym for Asynchoronous Transfer Mode, a switching protocol.

AWG (American wire gauge) The adopted standard wire sizes, such as, No. 12 wire, No. 14 wire, et cetera. The larger the gauge number of the wire, the smaller the wire. Therefore, a No. 14 wire is smaller than a No. 12 wire.

backbone Connection points in a network carrying messages between distributed LANs.

backup procedures The provisions made for the recovery of data files and program libraries and for restart or replacement of equipment after the occurrence of a system failure or of a disaster.

background A background task or program runs while the user is doing something else. The most common example is a print spooler program. Used in contrast to foreground.

background processing The ability to complete tasks in the background.

bandwidth The range of frequencies available for signaling. The difference expressed in Hertz between the lowest and highest frequencies of a band.

baud Unit of signaling speed. The speed in baud is the number of discreet conditions or events per second.

bioelectromagnetics The study of the interactions between biological systems and electromagnetic fields.

bit A contraction of binary digit. The smallest unit of information that a computer can hold. The value of a bit (1 or 0) represents a simple two-way choice, such as yes or no, on or off, positive or negative, something or nothing.

bps (bits per second) Unit of data transmission rate.

bridge A device used to connect LANs by forwarding packets addressed to other similar networks across connections at the Media Access Control Data-Link level. Routers, which operate at the protocol level, are also called bridges.

broadband A transmission system in which signals are encoded and modulated into different frequencies and then transmitted simultaneously with other signals.

broadcast   A LAN data transmission scheme in which data packets are heard by all stations on the network.

buffer   A temporary holding area of the computer's memory where information can be stored by one program or device and then read at a different rate by another. For example, a print buffer.

bus   A common connection. Networks that broadcast signals to all stations, such as Ethernet and ARCnet, are considered bus networks.

byte   A unit of information having eight bits.

CCITT (International Telegraph and Telephone Consultative Committee) International standards-setting organization.

CDMA (Code Division Multiple Access)   A digital cellular technology promoted by Qualcomm.

CDPD (Cellular Digital Packet Data)   An open protocol for transmitting data over the existing AMPS cellular infrastructure at 19.2K baud.

cellular   A system of reusing bandwidth by dividing a region into small cells, each with a stationary radio antenna. As a mobile user moves from cell to cell, the neighboring cell sites detect the changing signal strength and pass off control accordingly, instructing the mobile unit to change frequencies if necessary.

central processing unit (CPU)   The "brain" of the computer. The microprocessor performing the actual computations in machine language.

channel   An information transfer path within a system. Might also refer to the mechanism by which the path is effected. Also, a specific frequency used as a carrier for a particular service, e.g., an AM radio station, a TV channel.

character   Letter, numeral, punctuation or any other symbol contained in a message.

charge   The electrical property of matter responsible for creating electrical fields. Charge is either negative or positive.

chip   Slang for a silicon wafer imprinted with integrated circuits.

circuit   The arrangement of the wires in power lines, wall wiring, and electrical appliances.

client   In a client/server database system, the computer (usually a workstation) that makes service requests.

client/server   A network system design in which a processor or computer designated as a server (file server, database server, etcetera) provides services to other client processors or computers.

collision   A garbled transmission resulting from simultaneous transmissions by two or more workstations on the same network cable.

communications base station   A high-powered station where a dispatcher coordinates use of channels by permitting only one conversation at a time on each channel (often with dispatcher).

communication link   An electrical and logical connection between two devices. On a local area network, a communication link is the point-to-point path between sender and recipient.

communication program   A program that enables the computer to transmit data to and receive data from distant computers through the telephone system.

conduction   Relay of electrical charge through a circuit by exciting free electrons.

configuration   (1) The total combination of hardware components (central processing unit, video display device, keyboard, and peripheral devices) forming a computer system. (2) The software settings allowing various hardware components of a computer system to communicate with each other.

control program   A program designed to schedule and supervise the performance of data processing work by a computing system.

CRC (cyclical redundancy check)   Mathematical formula for generating a checksum for detecting errors in a received frame.

crosstalk   The unwanted transmission of a signal on a channel that interfaces with another adjacent channel. Signal interference created by emissions passing from one cable element to another.

cryptoanalysis   The steps and operations performed in converting messages (cipher) into plain text (clear) without initial knowledge of the key employed in the encryption algorithm.

cryptographic system   The documents, devices, equipment, and associated techniques that are used as a unit to provide a single means of encryption (enciphering or encoding).

cryptography   Transformation of plain text into coded form (encryption) or from coded form into plain text (decryption).

cryptology   The field that includes both cryptoanalysis and cryptography.

CSMA/CA (Collision Sense Multiple Access with Collision Avoidance) Access protocol used by LANs.

CSMA/CD (Collision Sense Multiple Access with Collision Detection) Access protocol used by Ethernet LANs.

current   The flow of electric charge through a power line or an electric wire. The current in a line is like the water flowing through a pipe. Currents produce magnetic fields.

data   Processable information with the associated documentation. The input that a program and its instructions perform on, and which determines the results of processing.

database   (1) A collection of information organized in a form that can be readily manipulated and sorted by a computer user. (2) Short for database management system.

database management system   A software system for organizing, storing, retrieving, analyzing, and modifying information in a database.

database server   A database server is the "back-end" processor that manages the database and fulfills database requests in a client/server database system.

datagram   One packet of information and associated delivery information routed through a packet switching network.

data integrity   Verified correspondence between the computer representation of information and the real-world events that the information represents. The condition of being whole, complete, accurate, and timely.

data security   Data security is the result achieved through implementing measures to protect data against unauthorized events leading to unintentional or intentional modification, destruction, or disclosure of data.

data storage   The preservation of data in various data media for direct use by the system.

decipher   To convert, by use of the appropriate key, cipher text (encoded, encrypted) into its equivalent plain text (clear).

decrypt   See decipher.

DECT (Digital European Cordless Telephone)   The 1.8-GHz digital standard rapidly being adopted in Europe.

diffuse infrared transmission   Transmission of infrared signals over a wide field of view such that the signal can be received any place in the room or area of coverage.

digital   A system based on discrete states, typically the binary conditions of on or off.

digital transmission   A communications system that passes information encoded as pulses. Baseband networks use digital transmissions, as do microcomputers.

direct current (dc)   An electrical current that does not change strength or direction over time, but remains steady. Batteries have direct current.

domain   The set of objects that currently can be directly accessed by a principal (subject).

Doppler effect   The change in the apparent length of a sound, light, or other wave as its source moves towards you and away again.

DSSS (direct sequence spread spectrum)   A spread-spectrum method that divides the available bandwidth into three or four subchannels.

eavesdropping   Unauthorized interception of data transmissions.

electric field (EF)   The forces that electric charges exert on other objects. The radiation from an electric source.

electricity   The effect of charged particles on the move or at rest.

electromagnetic field (EMF)   The field around an electric force containing an electric field and a magnetic field.

electromagnetic radiation   Energy sent out through space by varying electric and magnetic fields. Light, radio waves, microwaves, television signals, X-rays, and cosmic rays are all forms of electromagnetic radiation.

electromagnetic spectrum   A breakdown of electromagnetic fields according to their frequency and wavelength. The spectrum divides into extra low frequency (ELF) radiation, very low frequency (VLF) radiation, radio frequency (RF), and microwave radiation.

energy   The work associated with moving an electric charge in the presence of an electric field.

Ethernet   A local area network protocol developed by Xerox.

FCC (Federal Communications Commission)   A U.S. organization that controls the allocation of radio frequencies.

FHSS (frequency hopping spread spectrum)   A spread-spectrum methodology that divides the available bandwidth into a large number of small subchannels. The NICs and access points then jump around these subchannels in a pseudorandom manner.

field   The radiation emanating from an electric source. Also called an electromagnetic field.

file server   A computer that provides network stations with controlled access to shareable resources.

frequency   The number of waves per second, usually measured in cycles per second or Hertz.

frequency band   A range of frequencies used for similar services, e.g., the AM radio band.

gateway   A device that provides routing and protocol conversion among physically dissimilar networks and computers, e.g., LAN to host, LAN to LAN, X.25, and SNA gateways.

hacker   A computer enthusiast. Also, one who seeks to gain unauthorized access to computer systems.

handshaking   A dialogue between a user and a computer, a computer and another computer, or a program and another program. Used for identifying a user and authenticating his or her identity through a sequence of questions and answers based on information either previously stored in the computer or supplied to the computer by the initiator of the dialogue.

hardware   In computer terminology, the machinery that forms a computer system.

Hertz (Hz)   A measure of frequency or bandwidth. The same as cycles per second.

host computer   The computer that receives information from and sends data to terminals over telecommunication lines. The computer that is in control in a data communication network. The host computer might be a mainframe computer, minicomputer, or microcomputer.

hub   (1) A device used on certain network topologies that modifies transmission signals, allowing the network to be lengthened or expanded with additional workstations. The hub is the central device in a star topology. (2) A computer that receives messages from other computers, stores them, and routes them to other computer destinations.

IEEE (Institute of Electrical and Electronic Engineers) One of several groups whose members are drawn from industry and who attempt to establish industry standards. The IEEE 802 committee has published numerous definitive documents on local area network standards.

information Includes input, output, software, data, and all related documentation.

information pool Consists of data designated as accessible by authorized individuals.

infrared Heat radiation. Part of the spectrum of electromagnetic radiation emitted by objects.

insulation Reduction in the flow of electric current by certain materials (insulators).

interface A device or program that allows two systems or devices to communicate with each other. An interface provides a common boundary between the two systems, devices, or programs.

input/output (I/O) The process by which information is transferred between the computer's memory and its keyboard or peripheral devices.

IPX (Internetwork Packet Exchange) NetWare protocol that provides datagram delivery of messages.

ISM-band (industrial, scientific and medical band) The portion of the electromagnetic spectrum allocated to industrial, scientific, and medical wireless applications. The spectrum frequency allocations for ISM are 902–928 MHz, 2.400–2.483 GHz, and 5.700–5.900 GHz.

I/O device (input/output device) A device that transfers information into or out of a computer.

ITU (International Telecommunications Union) An international standards-setting organization.

light The part of the electromagnetic radiation that is visible to humans.

LLC (Logical-Link Control) Data-Link layer protocol governing transmission, a.k.a. the IEEE 802.2 standard.

local area network (LAN) A communications system using directly connected computers, printers, and hard disks allowing shared access to all resources on the network.

MAC (Media Access Control) Data-Link layer protocol governing access to the transmission medium.

magnetic field (MF) The force that a moving electric charge exerts on other moving charges. The field produced by an electric current.

MIB (management information base) A table defining the management information available to SNMP.

microcomputer A general term referring to a small computer having a microprocessor, a.k.a. personal computer.

microwave   Electromagnetic radiation with a frequency at the low end of radio frequency radiation. For example, radar. A form of nonionizing radiation.

multipath interference   Interference resulting from signals arriving at different times through multiple paths of different length (e.g., reflections).

NetBIOS (Network Basic Input Output System)   A PC networking software component introduced by IBM in 1984. It has become a standard interface allowing different makes of network hardware and software to exchange information.

NetWare   A network operating system from Novell.

network   A collection of interconnected, individually controlled computers, printers, and hard disks, including the hardware and software used to connect them.

network station   Any PC or other device connected to a network by means of a network interface board and some communication medium. A network station can be a workstation, bridge, or server.

NIC (network interface card)   Hardware installed in network devices that enables them to communicate on a network.

node   A point of interconnection to a network. Normally, a point at which a number of terminals are located.

ODI (Open Data-Link Interface)   A protocol that supports media- and protocol-independent communications by providing a standard interface allowing network layer protocols to share hardware without conflict.

ohm   Measure of electrical resistance.

operating system   Software that controls the internal operations (housekeeping chores) of a computer system. Operating systems are specific to the type of computer used.

OSI (Open Systems Interconnect)   A seven-layer model for data communications adopted by the International Standards Organization.

package   A generic term referring to any group of detailed computer programs necessary to achieve a general objective. For example, an accounts receivable package would include all programs necessary to record transactions in customers' accounts, produce customers' statements, etcetera.

packet   A group of bits transmitted as a whole on a network.

PCMCIA (Personal Computer Memory Card International Association)   Often used to refer to the standards that this group has released defining a credit-card-size module for personal computers and the software to operate this module.

PCS (Personal Communication Services)   A wireless environment that will provide subscribers with portable two-way calling regardless of location. PCS is the proposed next generation of wireless network services. It will provide voice communications services similar to today's cellular services as well as fax, e-mail, data, and image.

PDA (personal digital assistant)   A term often used by personal computer makers to describe the next-generation palmtop PCs that incorporate pen technology, wireless communications, and personal computer functionality.

peripheral   Any device used for input/output operations with the computer's central processing unit (CPU). Peripheral devices are typically connected to the microcomputer with special cabling and include such devices as modems and printers.

polling   A means of controlling devices on a line.

power   The rate at which work is done.

print server   A device and/or program that manages printers. Print service is often provided by the file server but can also be provided from a separate LAN PC.

processing   A systematic sequence of operations performed on data.

protocol   A set of characters at the beginning and end of a message that enables two computers to communicate with each other.

protons   Large positively charged particles in the nucleus of every atom.

radiation   Energy propagated through space in waves or particles. Some common forms of radiation are X-rays, microwaves, light, and radio waves.

repeater   Remote stations surrounding home station that retransmit on a different frequency for greater total range. For example, base transmits at frequency A. Repeater listens on frequency A and retransmits on frequency B. Mobile unit listens on frequency B. In the other direction, mobile unit transmits on frequency A, which is retransmitted at frequency B, to which the base listens.

router   A network layer device that connects two networks using the same networking protocol.

security   Protection of all those resources that the client uses to complete its mission.

serial port   Serial data communications port. On a workstation, the port that can be identified as COM1 or COM2 and usually conforms to RS-232. Many workstation serial ports use a 9-pin connector.

server   A network device that provides services to client stations. Servers include file servers, disk servers, and print servers.

session   Interval during which two computers maintain a connection.

SNMP (Simple Network Management Protocol)   Industry-standard protocol that makes it easy to describe objects and let network management get information about the objects and set values in the objects. Originally developed as part of TCP/IP.

software   Programs and routines to be loaded temporarily into a computer system. For example, compilers, utilities, operating system, and application programs.

spectrum   The entire band of available radio frequencies for a given application, authorized by the FCC.

SPX (Sequenced Packet Exchange)   NetWare protocol that allows two applications to communicate across the network. Uses IPX to deliver the messages, but guarantees delivery and order of messages.

system integrity   The behavior of a hardware/software system that does the right things. Further, it does these things right and does them when they are needed.

TCP/IP (Transmission Control Protocol/Internet Protocol)   Protocol suite developed for the Internet. TCP is the primary transport protocol, and IP is the network layer protocol.

TDMA (time division multiple access)   A technology that competes with CDMA. It is a method of digital transmission for wireless telecommunications systems that allows many users to simultaneously access a single radio frequency band without interference. TDMA is the wireless access method of choice for U.S. cellular systems.

tesla   A measure of magnetic fields.

Token Ring   IBM-developed LAN using both a physical and logical ring topology and token passing to regulate network access. Data travels in one direction around the ring, with each node taking data in on its uplink side and transmitting on its downlink side.

topology   The physical layout of the network cabling.

transceiver   Unit containing both transmitter and receiver.

unlicensed frequencies   Bands where no license is required, but users must be prepared to deal with interference from other users and equipment.

user   Used imprecisely to refer to the individual who is accountable for some identifiable set of activities in a computer system.

voltage   A measure of the electrical power or potential on a power line. Voltage is measured in volts (V).

watt   A unit of measurement of electrical power.

wavelength   The distance between two successive peaks of a wave.

wideband   A communications channel that has greater bandwidth than voice-grade lines.

wiretapping   Monitoring or recording data as it moves across a communications link. Also known as traffic analysis.

workstation   A desktop computer that performs local processing and accesses LAN services.

X-rays   A form of ionizing radiation known to break chemical bonds.

Selected Bibliography

Books

Bates, Regis J. "Bud." 1994. *Wireless Networked Communications: Concepts, Technology, and Implementation.* New York: Windcrest/McGraw-Hill.

Brodeur, Paul. 1989. *Currents of Death.* New York: Simon and Schuster.

Canadian Institute of Chartered Accountants. 1991. *Managing and Using Microcomputers.* Toronto: CICA.

Carlson, A. B. 1986. *Communication Systems—an introduction to signals and noise in electrical communication.* Third Edition. New York: McGraw-Hill.

Corrigan, Patrick H. and Aisling Guy. 1989. *Building Local Area Networks with Novell's NetWare.* Redwood City, CA: M & T Books.

Cronin, Daniel J. 1986. *Microcomputer Data Security.* New York: Prentice-Hall.

Davies, D.W. and W.L. Price. 1984. *Security for Computer Networks.* New York: John Wiley & Sons.

Davis, Peter T. 1994. *Complete LAN Security and Control.* New York: Windcrest/McGraw-Hill.

Davis, Peter T. and Craig R. McGuffin. 1994. *Teach Yourself NetWare in 14 Days.* Indianapolis, IN: SAMS Publishing.

Day, Michael. 1992. *Enterprise Series: Downsizing to NetWare.* Carmel, IN: New Riders Publishing.

Dixon, R. C. 1984. *Spread Spectrum Systems.* New York: John Wiley & Sons.

Fites, Philip, Peter Johnston and Martin Kratz. 1989. *The Computer Virus Crisis.* Toronto: Nelson Canada.

Hoffman, Lance J. 1990. *Rogue Programs: Viruses, Worms and Trojan Horses.* United States: Van Nostrand Reinhold.

Kane, Pamela. 1989. *V.I.R.U.S. Protection.* New York: Bantam Books.

Kosiur, Dave and Nancy E. H. Jones. 1992. *Macworld Networking Handbook*. San Mateo, CA: IDG Books.

Lundell, Allan. 1989. *Virus!*. Chicago: Contemporary Books.

McAfee, John and Colin Haynes. 1989. *Computer Viruses, Worms, Data Diddlers, Killer Programs and Other Threats to Your System*. New York: St. Martins Press.

Smith, Cyril W. and Simon Best. 1989. *Electromagnetic Man*. London: J.M. Dent & Sons Ltd.

Stallings, William. 1988. *Data and Computer Communications*. 2nd edition. New York: Macmillan.

Stallings, William. 1990. *The Business Guide to Local Area Networks*. Carmel, IN: Howard W. Sams & Company.

Stang, David J. 1990. *Computer Viruses*. Washington, DC: National Computer Security Association.

Sugarman, Ellen. 1992. *Warning: The Electricity Around You May Be Hazardous to Your Health*. New York: Fireside Books.

Articles and Publications

Allen, R. C. March 1991. Infralan wireless networks. *IEEE Document P802.11/91-33*.

Barron, J. J. June 1990. Want to catch some z's? *BYTE*. pp. 217–222.

Biba, K. J. June 1975. Integrity Considerations for Secure Computer Systems. *Report MTR-3153*. Bedford, MA: Mitre Corporation.

Blanc, J. L. September 1991. EEC policies and plans for wireless LANs. *Proceedings of the Wireless LANs conference*. London: IBC Technical Services Ltd.

Court, D. September 1991. The ERO and frequency allocations for radio LANs. *Proceedings of the Wireless LANs conference*. London: IBC Technical Services Ltd.

Datapro Research Corporation. 1985. *All About Local Area Networks and PABX Systems*. Delran, NJ: Datapro.

Deloitte & Touche. 1993. *Convergence: Destination Unknown*. Toronto.

Deloitte & Touche. 1993. *Welcome to the Untethered World : The Impact of Wireless Communications on Business in the 1990s*. Toronto.

Fay, G. April 1991. Closing the gap with wireless networks. *Communications*.

Freeburg, T. May 1991. What do you mean, radio? *IEEE Document P802.11/91-48*.

Gahan, C. September 1991. Infralan, a new approach to LAN flexibility. *Proceedings of the Wireless LANs conference*. London: IBC Technical Services Ltd.

Gunn, Angela. August 1993. Connecting Over the Networks. *PC Magazine*: 359–384.

Harper, R.M. Fall 1986. Internal Control of Microcomputers in Local Area Networks. *Journal of Information Systems*: 67–80.

Harper, R.M., A.H. Friedberg and M.J. Cerullo. Fall 1987. Local Area Networks: The PC Connection. *Journal of Accounting & EDP*: 4–11.

Hayes, V. September 1991. International Standardization. *Proceedings of the Wireless LANs conference.* London: IBC Technical Services Ltd.

Lessard, A. and M. Gerla. May 1988. Wireless communications in the automated factory environment. *IEEE Network.* Vol. 2, No. 3.; 64–69.

Marcus, M. J. 1985. Recent US regulatory decisions on civil uses of spread spectrum. *Proceedings of the IEEE Globecom.* New York: IEEE.: 16.6.1–3.

McMullen, Melanie. November 1992. LAN on the Run. *LAN Magazine*: 67–74.

Poor, A. October 15, 1991. Securing your LAN: Safe stations for networks. *PC Magazine.* Vol. 10, No. 17.: 215–288.

Rappaport, T. S. May 1989. Indoor radio communications for factories of the future. *IEEE Communications Magazine.*: 15–24.

Ratcliffe, M. September 10, 1991. How safe are wireless nets? *MacWeek.*: 1.

Rizzo, Tony. March 1991. Wireless LANs_Motorola Starts Breaking Down the Barriers to Acceptance. *Network Computing*: 10–12.

Rothberg, M. L. August 1990. IEEE 802 standards for local area networking. *Datapro Reports on Data Communications.* McGraw-Hill. C07-527-101.

Spencer, M. January 1990. Data Safety, special report. *MacWorld.*: 242–149.

Theodore, D. S. July 1991. LAN interconnect takes to the airwaves. *Data Communications.*: 83–89.

Touche Ross. 1993. *Mobilizing Your Business: The Impact of Wire-less Communications on Business in the 90s.* London.

Tuch, B. July/August 1993. Wireless Data Networking. *AT&T Technical Journal.* Vol. 72, No. 4, pp. 27–37.

Wittman, Art. June 1, 1994. Will Wireless Win the War? *Network Computing.* pp. 58–70.

Periodicals

Communications Technology
50 S. Steele Street
Suite 700
Denver, CO 80209-0000

Communications Week
PO Box 1094
Skokie, IL 60076

Communications World
1 Hallidie Place
Suite 600
San Francisco, CA 94102-2818

Computer Technology Review
924 Westwood Blvd.
Suite 650
Los Angeles, CA 90024-2927

Corporate Communications Report
112 E 31st Street
New York, NY 10016-6809

Data Communications Magazine
McGraw-Hill Inc.
1221 Avenue of the Americas
New York, NY 10020

EDN
275 Washington Street
Newton, MA 02158

Electronic Business
275 Washington Street
Newton, MA 02158-1611

Electronic Engineering Times
600 Community Drive
Manhasset, NY 11030-3847

Electronic News
7 East 12th Street
New York, NY 10003-4404

Electronics
10 Holland Drive
Hasbruck Heights, NJ 07604-3118

Engineering Times
1420 King Street
Alexandria, VA 22314-2750

Executive Report
3 Gateway Dr.
5th Floor
Pittsburgh, PA 15222-1004

Federal Communications Technology News
P.O. Box 435
Falls Church, VA 22046

Futurist
4916 St Elmo Avenue
Bethesda, MD 20814-6031

High Technology Business
270 Lafayette Street
Suite 705
New York, NY 10012-3327

Hospitals
211 E. Chicago Avenue
Suite 700
Chicago, IL 60611-2692

IEEE Spectrum
345 E. 47th Street
11th Floor
New York, NY 10017-2330

Industry Week
1100 Superior Avenue E.
Cleveland, OH 44114-2518

Information Week
600 Community Drive
Manhasset, NY 11030-3847

LAN Times
1900 O'Farrell Street, Suite 200
San Mateo, CA 94403

MacGuide Magazine
550 S. Wadsworth Blvd.
Suite 500
Lakewood, CO 80226-3121

MacIntosh Buyer's Guide
660 Beachland Blvd.
Vero Beach, FL 32963-1793

MacIntosh Hands On
52 Domino Drive
Concord, MA 01742-2817

MacIntosh News
600 Community Drive
Manhassett, NY 11030-0000

Macintosh Standard
604 Pine Drive
Brightwater, NY 11718-1728

MacTech Quarterly
290 SW 43rd Street
Renton, WA 98055-4936

MacTutor
1250 N. Lakeview
Suite O
Anaheim, CA 92807-1831

MacUser
950 Tower Lane
18th Floor
Foster City, CA 94404-2121

MacWeek
301 Howard Street
15th Floor
San Francisco, CA 94105-2241

MacWorld
501 2nd Street
Suite 600
San Francisco, CA 94107-1431

Mobile Executive Magazine
305 Broadway
Suite 305
New York, NY 10007

Mobile Office Magazine
21700 Oxnard Street
Suite 1600
Warner Woodland Hills, CA 91367

Modern Office Technology
1100 Superior Avenue E.
Cleveland, OH 44114-2518

National Report on Computers & Health
11300 Rockville Pike
Suite 1100
Rockville, MD 20852-3030

New Technology Week
529 14th Street NW
Suite 627
Washington, DC 20045-0001

Office Products Industry Report
301 N. Fairfax St.
Alexandria, VA 22314-2633

Open Systems Today
600 Community Drive
Manhasset, NY 11030-3875

PC Magazine
1 Park Avenue
New York, NY 10016-5802

PC Resource
80 Elm Street
Peterborough, NH 03458-1000

PC Week
800 Boylston Street
Boston, MA 02199-8088

PC World
501 2nd Street
Suite 600
San Francisco, CA 94107-1431

Pen Computing
7 Quadrini
Albany, NY 12208

Popular Science
2 Park Avenue
5th Floor, New York, NY 10016-5601

Radio Electronics
5008 Bi County Blvd.
Farmingdale, NY 11735-3918

Research & Development
1350 E. Touhy Avenue
Des Plaines, IL 60018-3303

Scientific American
415 Madison Avenue
New York, NY 10017-1179

Sound & Communications
25 Willowdale Avenue
Port Washington, NY 11050-3716

Telecommunications
685 Canton Street
Norwood, MA 02062-2610

Teleconnect
12 W. 21st Street
10th Floor
New York, NY 10010-6997

Telephone Switch Newsletter
5772 Bolsa Avenue
Suite K100
Huntington Beach, CA 92647-0000

Telephony Magazine
55 E. Jackson Blvd.
Suite 1100
Chicago, IL 60604-4188

Wireless Design and Development
Gordon Publications, Inc. (division of Cahners)
P.O. Box 650
Morris Plains, NJ 07950

Index

A

AIRONET Wireless Communications Inc., 167-173, 208, 211
Alps Electric, Inc., 173-174, 208
amplitude modulation (AM), 49
amps, 123
applications, 7-10, 14-22
  agricultural industry, 89
  braille display units, 96
  courthouse terminals, 97
  data acquisition equipment, 90
  disaster recovery, 98-99
  electronic mail, 97
  engine link to diagnostic computer, 90-91
  factory control, 89-90
  financial services industry, 107-110
  flight control, 93
  food industry, 89
  food services industry, 110-112
  health care industry, 101-106
  industrial, 88-92
  industry, 101-117
  inventory control, 91
  metalworking equipment, 91
  oceanographic research buoys, 93
  office sharing, 97
  office, 93-98
  overhead crane control, 91
  PC to wired network, 96
  petroleum industry, 89
  plastics factory, 91
  printer sharing, 96
  programmable road signs, 93
  remote control, 96
  retail industry, 112-117
  robotics, 91-92
  scoreboards, 92
  services industry, 110-112
  speed measurement, 93
  summary of wireless, 116
  transportation industry, 92-93
  weighing scales, 91
  weight measurement, 93
AT&T Global Information Solutions, 174-175, 208, 211
automatic guided vehicle (AGV), 88

B

bandwidth, 12, 15, 47
Black Box Corporation, 175-177, 208, 211
bounded media, 27
bridges, 39
broadband, 11
bus topology, 32-33, 55

C

Cabletron Systems, 177
California Microwave, 177
carrier sense multiple access (CSMA), 35
cellular telephone, 16-17
client nodes, 32
collision avoidance (CA), 35
collision detection (CD), 35
Commodity Exchange Inc. (COMEX), 107
communication protocols, 36-38
communications
  equipment costs, 48-49
  information-carrying capacity, 47
  infrared, 16
  meteor burst, 18
  mobile satellite, 19-22
  modulation, 49

communications, *continued*
  range of transmission, 46-47
  regulation, 47-48
  wireless (*see* wireless data communications)
Communications Research and
    Development Corporation, 178-180, 208
computer-aided design (CAD), 88-89
computer-aided engineering (CAE), 88-89
computer-aided manufacturing (CAM), 88-89
computer technology, 8-9
contention methods, 35-36
Cylink Corporation, 180-181, 214

D
database servers, 31
deterministic methods, 36
Digital Ocean Inc., 181
direct sequence spread spectrum (DSSS), 50
disaster recovery, 98-99

E
electricity, 122-123
  measuring, 123
electromagnetic fields, 121-122, 124
electromagnetic radiation, 44-50
electromagnetic spectrum, 45, 124-125
electromagnetic wave, 45
EMF emitters, 123
encryption, 142-145
European Telecommunications Standards
    Institute (ETSI), 52, 64, 67

F
Federal Communications Commission
    (FCC), 47
file servers, 30-31
financial services industry, 107-110
FM sideband, 22
FM squared, 21-22
food services industry, 110-112
frequency, 10, 12, 15
  used for wireless LANs, 51-52
frequency bands, 215-216
frequency-hopping spread spectrum
    (FHSS), 50
frequency modulation (FM), 49

G
gateways, 39
Granite Communications, 181-183, 209, 211

H
health and safety, 24, 121-126
  electricity and, 122-123
  electromagnetic fields, 121-122, 124
health care industry, 101-106
Hewlett-Packard Corporation, 183-184

I
IBM Corporation, 184-185
industrial applications, 88-92
industry applications, 101-117
industry standard architecture (ISA), 61
INFRALAN Technologies, Inc., 184, 211
infrared communications, 16
Institute of Electrical and Electronic
    Engineers (IEEE), 61, 64, 221-223
International Electrotechnical Commission
    (IEC), 64
International Frequency Regulatory Board, 47
International Standards Organisation (ISO), 37
Internet newsgroup, 223-228

L
Laser Communications, Inc., 185-186, 214
light waves, 46
line-of-sight infrared transmission, 10
local area network (LAN), 25-40
  communication protocols, 36-38
  definition/description, 26-27
  integrating with wireless, 75-80
  internetworking devices, 38-39
  media connections, 32-36
  nodes, 27-32
  vs. wireless, 73-75, 87
  wired and wireless connections, 11
  wireless (*see* wireless LANs)
  wiring costs, 80

M
management, 147-156
  asset, 155-156
  fault, 153-154
  implementation of wireless LAN, 147-152
  performance, 154-155
  pilot projects, 148-152
media
  bounded, 27
  connecting, 32-36
  contention access control, 35-36
  controlling access to, 34-36
  deterministic access control, 36
  unbounded, 27

media access control (MAC), 65-67
mesh topology, 34
meteor burst communications, 18
metropolitan area network (MAN), 27
Microwave Radio Corporation, 186-188, 214
microwaves, 18-19, 46
Mitsubishi Electronics America Inc., 188
mobile radio, 17-18
mobile satellite communications, 19-22
modems, 16-17
modulation, 49
Motorola Wireless Data Group, 188-190, 212

N

network interface card (NIC), 28
networks and networking
  local area (LAN), 25-40
  metropolitan area (MAN), 27
  roles of, 28-29
  topologies, 32-34
  wide area (WAN), 27
  wireless (see wireless LANs)
nodes, 27-32
  client, 32
  services offered by, 29-30
Norand Data Systems Ltd., 190-191

O

O'Neill Connectivity, Inc., 191-192, 212
office applications, 93-98
open systems interconnect (OSI), 37, 62-64
organizations, 217-219

P

Persoft, Inc., 192-193, 212
personal communications service (PCS), 52
personal digital assistant (PDA), 103
Photonics, 193, 212
print servers, 31
private branch exchange (PBX), 98
probabilistic techniques, 35
programmable logic controller (PLC), 88
protocols, 36-38
Proxim, Inc., 193-196, 209

R

radiation, electromagnetic, 44-50
radio
  mobile, 17-18
  spread-spectrum, 15-16
radio spectrum, 48
radio waves, 46, 47

receivers, 43
repeaters, 39
retail industry, 112-117
ring topology, 33, 55
routers, 39

S

safety (see health and safety)
satellite communications, mobile, 19-22
security, 24, 127-145
  eavesdropping, 133-134
  encryption, 142-145
  equipment theft, 133
  features in NOS, 141-142
  goals, 130-131
  interception of messages, 133-134
  interference, 139
  masquerading, 138
  measures to taken, 132, 140-145
  modifying/substituting data, 136-138
  necessity of, 129-132
  shielding radiation, 140-141
  threats/risks, 131-132
  wireless LANs and, 132-139
servers
  database, 31
  file, 30-31
  print, 31
  specialty, 31
services industry, 110-112
Sharp Electronics, 196
shielding, 128, 140-141
shortwaves, 46
small computer system interface (SCSI),
  61
Solectek, 196-197, 209, 212-213
sound waves, 42
South Hills Datacomm, 197-199, 209, 213
Spectrix Corporation, 209, 199
spread-spectrum radio, 15-16
spread-spectrum technology (SST), 48, 50
standards, 23, 59-69
  affected OSI layers, 62-63
  current, 64-67
  ETSI, 67
  future development of, 67-68
  IEEE 802.11, 65
  IEEE P802.11, 221-223
  impact on current wireless LANs, 68-69
  MAC level, 65-67
  necessity of, 60-61
star topology, 34, 55

T

telephone, cellular, 16-17
Telesystems, 209-210, 213
topologies, 32-34
  bus, 32-33, 55
  mesh, 34
  ring, 33, 55
  star, 34, 55
  tree, 34
transmission control protocol/internet pro-
    tocol (TCP/IP), 61
transmitters, 43
transportation industry, 92-93
tree topology, 34

U

unbounded media, 27

V

vendors, 167-214
volts, 123
VSAT technology, 19-22

W

warehousing, 115, 117
watts, 123
wavelength, 44
waves, 44-50
  electromagnetic, 45
  light, 46-47
  microwave, 18-19, 46
  need for regulation of, 47-48
  radio, 46, 47
  range of transmission, 46-47
  shortwave, 46
Wi-LAN, Inc., 199-201, 214
wide area network (WAN), 27
Windata, Inc., 201-202, 213-214
wireless data communications
  advantages, 11-13
  basics of, 41-43
  benefits, 6

costs, 6
  key components, 43
  market for, 2-8
  projected growth in U.S., 5
wireless LANs
  advantages, 11-13, 87
  applications (*see* applications)
  architecture, 52-58
  buildings with hazardous materials, 79
  buildings with hostile/dangerous environ-
    ments, 79
  buildings with sealed rooms, 79
  connections and connecting, 11
    remote nodes, 79
    wireless nodes to wired LAN, 54
  considerations, 99-100
  costs, 13
  factory locations of, 78
  frequencies used for, 51-52
  future, 9-10, 159-163
  historical buildings and, 77-78
  integrating with wired, 75-80
  management, 147-156
  market, 4
  market growth, 6
  mobility of, 81-82
  pen-based systems, 105
  replacing point-to-point link, 53
  revenues, 5
  selection checklist, 100
  standalone nodes, 55-58
  standards, 59-69
  temporary locations, 80-81
  uses for, 73-82
  vs. wired, 39-40, 73-75, 87
wireless technology, 41-58
  failure of, 160
  future, 23-24
  promise of, 160-162

X

Xircom, Inc., 203-204

ABOUT THE AUTHORS

In his 19 years in information systems, PETER DAVIS worked in data processing in large-scale installations in the financial and government sectors, where he was involved in the development and implementation of applications and specification of requirements. Most recently, he worked as director of information systems audit for the Office of the Provincial Auditor (Ontario). In addition, Peter was a principal in an international public accounting firm's information systems audit practice, and has acted as the Canadian representative for a U.S. company specializing in the manufacture and integration of communications products. Peter is now principal of Peter Davis+Associates, a training and consulting firm specializing in the security, audit, and control of information systems.

He is the author of *Complete LAN Security and Control* (Windcrest/McGraw-Hill, 1994) and co-author of two other books. Peter also is an internationally known speaker on security and audit, frequently speaking at local user meetings and international conferences sponsored by professional organizations and industry groups. In addition, he has had numerous articles published on security and audit. Peter is a member of the international committee formed to develop Generally Accepted System Security Principles (GSSP). He also is an Advisory Council member for the Computer Security Institute.

Peter Davis received his Bachelor of Commerce (B. Comm) degree from Carleton University. He also is a Certified Management Accountant (CMA), Certified Information Systems Auditor (CISA), Certified Systems Professional (CSP), Certified Data Processor (CDP), Certified Computing Professional (CCP), Certified Information Systems Security Professional (CISSP) and Certified Novell Administrator v3.11 (CNA). He currently lives in Toronto, Ontario, with his wife and daughter.

CRAIG MCGUFFIN has more than 13 years of experience in the computer system implementation, controls, and security. He holds a Chartered Accountant designation and has a background in computer science obtained through his Bachelor of Mathematics (Honours) from the University of Waterloo.

He is currently the principal of C. R. McGuffin Consulting Services, a Toronto-based firm that helps its clients manage and control today's computer technology. During his years as an independent consultant, and as a senior manager in an international public accounting firm, he assisted clients with all sizes and types of computer environments, ranging from large multimainframe installations to interconnected minicomputers and local area networks distributed across the continent.

Craig also devotes time to helping businesspeople better understand and use computer and network technology. He is the co-author of an instructional book on Novell NetWare administration, and he has developed and delivered many computer systems-related courses. These include university courses required for a professional designation, as well as a number of customized multiday training programs for computer security professionals. Craig is also an international speaker on the use of computer technology, controls, and security through numerous television appearances, articles published in newspapers and business periodicals, and through public speaking engagements at a variety of conferences sponsored by professional organizations and industry groups. Craig can be reached directly via electronic mail through the Internet at CRMcGuffin@CRMcG.COM, or through CompuServe at 71151,250.